FUEL ECONOMY OF THE GASOLINE ENGINE

FUEL ECONOMY OF THE GASOLINE ENGINE

Fuel, Lubricant and Other Effects

Edited by

D. R. BLACKMORE and A. THOMAS

Shell Research Limited, Thornton Research Centre, Chester, United Kingdom

Contributors

W. S. Affleck, J. Atkinson, A. G. Bell,
B. Bull, R. Burt, B. D. Caddock,
G. A. Harrow, A. J. Humphrys,
E. L. Padmore, O. Postle, I. C. H. Robinson
and G. B. Toft
(all of *Shell Research Limited, Thornton Research Centre, Chester*)

A HALSTED PRESS BOOK

JOHN WILEY & SONS
New York

First published in Great Britain 1977 by
The Macmillan Press Ltd

Published in the U.S.A. by
Halsted Press, a Division of
John Wiley & Sons, Inc.
New York

Printed in Great Britain

Library of Congress Cataloging in Publication Data

Main entry under title:

Fuel economy of the gasoline engine.

"A Halsted Press book."
Includes bibliographical references and index.
1. Internal combustion engines, spark ignition – Fuel consumption. I. Blackmore, David Richard, 1938– II. Thomas, Alun, 1925– III. Affleck, W. S.
TJ789.F78 1977 621.43'4 77–3916
ISBN 0 470–99132–1

Contents

Preface

The fuel economy of the gasoline engine has suddenly become important. This importance dates from the 1973 October War in the Middle East and the subsequent very large increase in the price of crude oil imposed by the OPEC countries. Although the dating of this increase will be precisely marked in history, the underlying reasons extend over a longer period and can be clearly seen from figure 0.1 (published in 1972!) of the ratio of crude oil reserves to production rate. From 1949 onwards there is an inexorable decline in this ratio, which points very clearly towards a severe shortage in the long term. The October War and its influence on oil pricing served merely to bring us face to face with the background reality. Whether it be a physical shortage of crude oil, a high price to be paid by the customer or an extra burden on the national balance of payments of a specific country, the underlying cause is the expectation, based on the facts in the graph, that the demand for crude oil is going to exceed the supply greatly. (Production has levelled off since 1973, but the reserves/production ratio for 1975 was down to 30.5.)

There is therefore a very good reason to look hard at the efficiency with which oil resources are being used. In western Europe roughly one-quarter of crude oil is used in the transport sector, and of that 60% is used in private cars. The fraction of oil used for transport is likely to increase rather than to diminish as other fuels (notably coal and gas) replace oil in the industrial and domestic sectors, so that the fuel economy of the gasoline engine will grow in importance.

One must, however, remember that fuel economy is only one of the criteria by which the performance of a fuel–engine combination will be assessed. If one looks back at the development of the car engine, one sees an initial period of about thirty years from 1885, when the first spark-ignition engined car ran on the road, to about 1915, at the beginning of the First World War, during which cars were being developed to be reliable and to operate over a whole range of driving conditions. 'Driveability', one might say, was the predominant objective during that period. Then, under the influence of wartime pressures, power per unit weight became increasingly important. An era of 50 years, from 1915 to about 1965, ensued, during which the criterion of increased power per unit weight became the main theme of engine development. The Wankel engine would have been a logical successor in this era, but

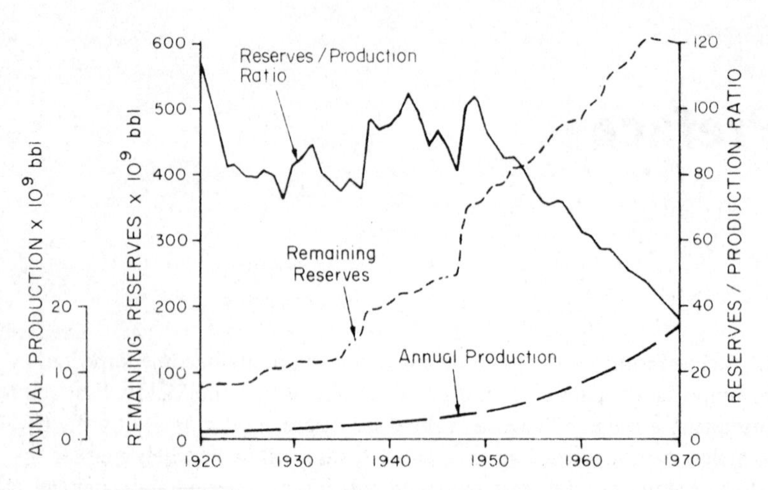

Figure 0.1 World production, remaining reserves and reserves/production ratio (from H. R. Warman. The future of oil. *Geograph. J.* (September 1972))

a new requirement emerged that made itself manifest in the 1960s: that of low emissions. This led to nearly a decade during which the main pressure in the evolution of the car became that of reducing pollution. And so we come to the October War in 1973 and a new and urgent emphasis on fuel economy. As successive objectives have become established the earlier ones have still remained, and now all four – good driveability, high power-to-weight ratio, low emissions and good fuel economy – have become cumulative and simultaneous requirements. Thus, engines of the future need to show all these four characteristics.

This book has been written for several reasons. First, although there are many excellent treatises which touch on the fuel economy of the gasoline engine, the subject is usually in a subordinate position under various headings, and information has to be extracted in a somewhat piecemeal fashion: here we have tried to marshal information on fuel economy in a more coherent form. Second, we have attempted a more explicit discussion of the way in which the requirement of fuel economy interacts with other desirable features, so that where a conflict exists the points to be reconciled can be more clearly recognized. Third, it has become clear that our techniques for measuring fuel economy need careful scrutiny in view of the increased load that they now are required to bear. We therefore include an extensive review of this subject.

Throughout the book, therefore, runs the theme of compromise between conflicting requirements. Only in the final chapter, 'Mileage Marathons', and in a somewhat different vein, is full rein given to the achievement of economy above all else. Described there is the evolution of a competition which is primarily a sporting and social event but which we think holds much that will be of considerable technical interest to a wide spectrum of readers.

The book started its life as an internal report for the Shell group of companies, but its initial reception has encouraged us to develop it for submission to a wider audience. It has been written from the point of view of the supplier of fuels and lubricants, with the emphasis mainly on fuels. Even so, this has entailed a generous treatment of engine factors. Our intention has been to provide a source of information, but a critique rather than a catalogue: each author has produced not only an account of his subject in factual terms, but also, where practicable, a considered opinion about it. No attempt has been made to harmonize the style of writing nor to eliminate the slight overlapping or even repetition of subject matter in one or two places, for this is in a sense a collection of essays by individuals.

Mention must be made of the systems of units used in this book. An attempt has been made to give all quantities in SI units, but it is recognized that many workers in this field are more familiar with the traditional units of their country. Where the results were originally presented in units other than SI, these units have been retained, but conversions are given where they are judged to be of particular use. They are not given, for example, where a table or figure is mainly used for internal comparisons. For those who require more extensive conversions to SI units, a set of useful factors will be found in appendix F.2.

It is a pleasure to acknowledge the assistance and encouragement of other colleagues, particularly Mr A. Birks and Mr A. Crisp, in the preparation of the book, and we are grateful to Shell Research Ltd for permission to publish it. Our thanks are also due to Mr D. S. Longbottom for preparing the index. A full list of acknowledgements is given overleaf.

Chester, 1977

D. R. B.
A. T.

Acknowledgements

The authors and publishers wish to thank the following individuals and organisations who have kindly assisted in supplying information and granting permission to use copyright material:

American Chemical Society (*Preprints*: Publications of the Division of Petroleum Chemistry Inc.)
American Petroleum Institute
Dr. W. M. Brehob and Dr. J. A. Robison
Champion Sparking Plug Company Ltd.
Institution of Mechanical Engineers
McGraw-Hill Book Company
The Editor of *Motor*
Motor Industry Research Association
The MIT Press, Cambridge, Massachusetts (figures 2.2 to 2.10, and 2.15)
Rand Corporation
Ricardo and Co., Engineers (1927) Ltd.
Royal Geographical Society
The Society of Automotive Engineers Inc. (figures 2.11 to 2.14, 2.16 to 2.21, 3.9 to 3.11, 5.2, 6.11, 8.1 to 8.6, 8.8, 8.9, 9.5, 10.2 to 10.4, 11.1 and 11.2, and tables 9.4, 9.11, 9.12 and 11.1 to 11.16)
H. R. Warman

1 Introduction

D. R. BLACKMORE

On the subject of motor car fuel consumption, apparently, 'everybody is an expert'. In the aftermath of the oil supply crisis in the autumn of 1973, public consciousness of this topic is again high, and the technical world is now engaged in this subject area at a new level of priority. Yet the information available on the factors that govern fuel consumption is, in general, surprisingly poor and more so in Europe than in the US. For instance, the measurement of vehicle fuel economy in relation to the way the motorist drives is a surprisingly difficult thing to do repeatably, because of the variety of ways and weather conditions in which motorists actually drive, combined with the rather poorly researched experimental techniques currently available. Methods for improving the fuel economy of a vehicle without materially affecting its performance in other respects are at an early stage of development, and much work needs to be done before the most cost-effective solutions can emerge.

This book has been written to help fill the information gap. It has deliberately been written from the viewpoint of an oil company's product research activities. The chapters that follow deal with the effect of the properties of oil products (mainly gasoline but also to a lesser extent lubricants) on fuel economy and also with related topics (e.g. the effect of mixture quality, vehicle maintenance and emission-control devices).

The direct quantitative effect on fuel economy of differences in oil product properties is not huge, especially when compared with effects obtainable from changes in engine and vehicle design, customer choice (engine capacity, size and weight of vehicle) and driving patterns (speed limits, traffic control, road design, acceleration rates, etc.). Yet fuel and lubricant effects are significant; they are worth studying, and the options are worth opening up for consideration. Moreover, indirect effects can be important in an era of evolving design (e.g. with respect to emission control and safety features), and it is distinctly possible that significant gains may result from novel lines of research into oil products. The chapters that follow are therefore intended to be open ended to some degree and to point to areas of ignorance, half-knowledge or potential opportunity.

There are four principal groups of people who need good information on this subject, and not surprisingly each group requires somewhat different information.

(1) *Customers* The car owner needs to know how to improve the fuel economy of the vehicle in service, both in choosing the fuel and lubricants and also in making decisions concerning effective maintenance. The car purchaser needs to know the relative fuel economy of the cars available and in more detail how this might vary with driving pattern.

(2) *Motor manufacturers* The engine designer needs to know what options are required and how these are limited (or are potentially extended) by the fuels and lubricants that may emerge in the future; since fuel economy is affected by factors other than engine characteristics, the transmission, tyres, vehicle drag (styling), accessories, weight, etc., all need to be chosen with respect to their effect on fuel consumption and on the overall cost to the consumer of running the car.

(3) *Oil companies* The effects of changes in fuel and lubricant design need to be known in more detail by the research, manufacturing and marketing sectors so that the consequences of such changes may be better understood and, if need be, may be planned for with sufficient lead time.

(4) *Government authorities and agencies* Both national and international requirements need to be catered for, and it is essential that decisions relating to fuel economy are made with the best available information. The authorities not only have to look at questions of national economy (balance of payments, energy consumption, etc.) but also within each country usually have a mandate to look after the customer, the motor industry and the oil industry. So theirs is a difficult task, particularly since it concerns a number of different government agencies in most countries.

As a background to this subject, we present for the first time a brief analysis of the current gasoline scene in the UK and the events of the preceding years as they pertain to motor cars. Of particular interest are the effects on gasoline usage of the events of 1973 and their aftermath, the data for which are now just beginning to emerge. We have chosen the UK as an example because the basic data are readily available to us (see tables 1.1–1.3) and because it is typical of a European country (as opposed to the US). It should be added that significant differences do exist within Europe and with other developed countries (e.g. Japan, Australia, South Africa, etc.), but similar data for these countries could reasonably easily be put together.

Of even greater interest than the events of the past decade are those of the next one or two decades. We do not attempt here to discuss in detail such a difficult field. Suffice it to say that a very thorough forecast of vehicles and traffic has recently been made[7] for Great Britain, based on extrapolations of data on car ownership per head of population; in spite of suggested low rates of growth in the national economy and substantial fuel price increases, the forecast predicts a 70–80% increase in traffic volume between 1973 and 2000. Such a figure indicates yet again the very high priority that an industrially developed society puts on vehicle ownership and use.

Table 1.1 *Gasoline deliveries and end usage*

Year	Total gasoline delivered in the UK[a], 10^6 tons (refs. 1, 2)	Total gasoline delivered in GB, 10^6 tons (ref. 1)	Gasoline for cars and motorcycles delivered in the UK, 10^6 tons (refs. 1, 2)	Gasoline for cars and motorcycles delivered in GB, 10^6 tons (refs. 1, 2)	Motorcycle gasoline usage in GB[b], 10^6 tons	Gasoline for cars delivered in GB[c], 10^6 tons
1975				(12.63)[d]	(0.13)	(12.5)
74	16.223	15.817	13.400	13.065	0.129	12.936
73	16.659	16.237	13.755	13.407	0.129	13.278
72	15.648	15.248	12.820	12.492	0.122	12.370
71	14.727	14.339	11.940	11.625	0.131	11.494
1970	14.010	13.639	11.250	10.952	0.134	10.818
69	13.231	12.891	10.460	10.191	0.144	10.047
68	12.808	12.475	10.070	9.808	0.162	9.646
67	12.084	11.770	9.360	9.117	0.180	8.937
66	11.322	11.030	8.620	8.398	0.206	8.192
65	10.739	10.467	7.951	7.750	0.230	7.520
64	10.012	9.755	7.190	7.005	0.260	6.745
63	9.043		6.494	(6.327)[e]	0.263	6.064
62	8.565		5.763	(5.615)	0.299	5.316
61	8.143		5.300	(5.164)	0.335	4.829
1960	7.625		4.723	(4.602)	0.347	4.255
59	7.124		4.212	(4.104)	0.337	3.767
58	6.623		3.636	(3.543)	0.288	3.255
57	5.745		3.040	(2.962)	0.287	2.675
56	6.323		3.140	(3.060)	0.255	2.805
55	6.240		3.000	(2.923)	0.259	2.664
54	5.922		2.700	(2.631)	0.237	2.394
53	5.740		2.632	(2.564)	0.231	2.333
52	5.441		(2.50)[f]	(2.44)	0.207	2.23

Table 1.1 (*continued*)

Year	Total gasoline delivered in the UK[a], 10^6 tons (refs. 1, 2)	Total gasoline delivered in GB, 10^6 tons (ref. 1)	Gasoline for cars and motorcycles delivered in the UK, 10^6 tons (refs. 1, 2)	Gasoline for cars and motorcycles delivered in GB, 10^6 tons (refs. 1, 2)	Motorcycle gasoline usage in GB[b], 10^6 tons	Gasoline for cars delivered in GB[c], 10^6 tons
1951	5.454		(2.51)[f]	(2.45)	0.192	2.26
1950	5.195		(2.39)[f]	(2.33)	0.150	2.18
49	4.671		(2.15)[f]	(2.09)	0.106	1.98
48	4.264		(1.96)[f]	(1.91)		
47	–					
46	–					
1938	4.83		(2.22)[f]	(2.17)	0.108	2.06

The bracketed values are estimates made by ourselves.
[a] UK ≡ GB + N. Ireland.
[b] Calculated from numbers of motorcycles and their annual distances driven given in references 3 and 4, assuming an average fuel economy of 60 mile/gal.
[c] Calculated from the fifth and sixth columns of this table.
[d] Assuming that the 3.3% decrease in the first 6 months of 1975 compared with 1974 applies for the whole year.
[e] Assuming that the 1964 ratio for total delivery in GB to total delivery in the UK applies for all preceding years.
[f] Assuming that gasoline deliveries for cars and motorcycles are 0.46 of the total deliveries (i.e. the 1953 value).
1 ton = 1016 kg ≈ 1364 l of gasoline.

Table 1.2 Gasoline retail prices, tax and cost to the average worker in time worked to buy one gallon

Year	Typical retail price of premium gasoline (inner region, 2000 gal drop), p/gal (reference 5)	UK tax per gallon, p (reference 5)	Average wage of a manual male worker, p/h (reference 6)	Time worked to buy a gallon of gasoline, min
1975 (Oct.)	68.5	22.5 + 13.7 VAT	135.6	30.3
75 (April)	68.5	22.5 + 13.7 VAT	122.9	33.4
74 (Oct.)	51.5	22.5 + 3.8 VAT	107.8	28.7
74 (April)	47.5	22.5	94.0	30.3
73 (Oct.)	37.2	22.5	89.7	24.8
73 (April)	35.2	22.5	84.1	25.1
72 (Oct.)	35.2	22.5	79.6	26.5
72 (April)	34.3	22.5	73.8	27.8
71 (Oct.)	34.0	22.5	69.2	29.5
1970 (Oct.)	32.4	22.5	61.4	31.7
69 (Oct.)	31.9		53.4	35.8
68 (Oct.)	31.5	20	49.6	38.1
67 (Oct.)	28.0	18	46.3	36.3
66 (Oct.)	27.0	18	44.1	36.7
65 (Oct.)	26.5	16	41.7	38.1
64 (Oct.)	27.0	14	38.0	42.6
63 (Oct.)	24.0	14	35.2	40.9
62 (Oct.)	24.5	13.5	33.7	43.6
61 (Oct.)	24.5	12.5	32.4	45.4
1960 (Oct.)	23.0		30.3	45.5
59 (Oct.)	23.0		27.9	49.5
58 (Oct.)	23.5		26.9	52.4
57 (Oct.)	23.5	12	26.1	54.0
56 (Oct.)	30.0		24.5	73.5
55 (Oct.)	23.0	12.5	22.8	60.5
54 (Oct.)	22.5		21.1	64.0
53 (Oct.)	24.5	12.5	19.8	74.2
52 (Oct.)	23.0	12.5	18.7	73.8
51 (Oct.)	22.0	9.5	17.4	69.0
1950 (Oct.)	15.5	7.5	15.8	58.9
49 (Oct.)	11.0	3.8	15.3	43.1
48 (Oct.)	10.5		14.8	42.7
47 (Oct.)	9.5		13.8	41.5
46 (Oct.)	9.0		12.7	42.6
1938 (Oct.)	8.0	3.8	7.23	66.4

Table 1.3 GB annual car gasoline usage, distance travelled and fuel economy

Year	No. of cars registered, million (reference 3)	No. of motor cycles registered, million (reference 3)	Gasoline per car per year		Total car + taxi miles x 10^9 (reference 4)	No. of taxis[c] registered, thousand (reference 3)	Taxi miles[d] x 10^9	Total car miles x 10^9	Average annual miles per car[e]	Average car fuel economy[f], mile/gal (l/100 km)	
			ton[a]	gal[b]							
1975	(13.65)	(1.0)	0.916	275							
74	13.639	(1.0)	0.948	285	117.41	(28.0)	(1.05)	116.36	8530	30.0	(9.41)
73	13.497	1.006	0.984	295	120.88	28.8	1.08	119.80	8880	30.1	(9.38)
72	12.717	0.982	0.973	292	114.50	28.0	1.05	113.45	8920	30.6	(9.23)
71	12.062	1.021	0.953	286	108.07	28.1	1.06	107.01	8870	31.0	(9.11)
1970	11.515	1.048	0.939	282	100.25	25.2	0.95	99.30	8620	30.6	(9.23)
69	11.227	1.127	0.895	268	93.82	22.5	0.84	92.98	8280	30.8	(9.17)
68	10.816	1.228	0.892	267	90.41	19.7	0.74	89.67	8290	31.0	(9.11)
67	10.303	1.350	0.867	260	85.17	15.4	0.58	84.59	8210	31.6	(8.94)
66	9.513	1.406	0.862	259	78.61	15.3	0.57	78.04	8200	31.7	(8.91)
65	8.917	1.612	0.844	253	71.96	14.8	0.56	71.40	8000	31.6	(8.94)
64	8.247	1.741	0.819	246	65.67	14.6	0.55	65.12	7900	32.1	(8.78)
63	7.375	1.755	0.822	247	56.82	14.4	0.54	56.28	7630	30.9	(9.14)
62	6.556	1.779	0.812	244	52.04		(0.5)	51.54	7860	32.2	(8.77)
61	5.979	1.790	0.808	242	47.79		(0.5)	47.29	7910	32.7	(8.64)

1960	5.526	1.796	0.770	231	42.29		(0.5)	41.79	7560	32.7	(8.64)
59	4.966	1.679	0.759	228	38.64		(0.5)	38.14	7680	33.7	(8.38)
58	4.548	1.475	0.716	215	34.41		(0.5)	33.91	7460	34.7	(8.14)
57	4.186	1.431	0.639	192	28.11		(0.5)	27.61	6600	34.4	(8.21)
56	3.887	1.290	0.722	216	28.73		(0.5)	28.23	7260	33.6	(8.40)
55	3.525	1.221	0.756	227	26.29		(0.5)	25.79	7320	32.2	(8.77)
54	3.099	1.108	0.773	232	23.08		(0.5)	22.58	7290	31.4	(8.99)
53	2.761	1.009	0.845	253	20.73		(0.5)	20.23	7330	28.9	(9.77)
1952	2.508	0.922	0.889	267	19.02		(0.5)	18.52	7380	27.6	(10.23)
51	2.380	0.823	0.950	285	18.22		(0.5)	17.72	7450	26.1	(10.82)
1950	2.258	0.729	0.965	290	15.93		(0.5)	15.43	6830	23.6	(11.97)
49	2.131	0.635	0.929	279	12.60		(0.5)	12.10	5680	20.4	(13.84)
48	1.961	0.543									
47	1.944	0.514									
46	1.770	0.449									
1938	1.944	0.444	1.06	318	17.27		(0.5)	16.77	8630	27.1	(10.42)

The bracketed values are estimates or calculations made by ourselves.
[a]Calculated from the seventh column of table 1 and the second column of this table.
[b]Assuming 1 ton is equivalent to 300 gal.
[c]Public transport vehicles of less than 8 seats.
[d]Assuming 25 000 mile/year per taxi and assuming that one-third of all taxis are privately registered.
[e]Calculated from the second and ninth columns of this table.
[f]Calculated from the fifth and tenth columns of this table.
1 ton gasoline $\approx$ 1364 l; 1 gal = 4.546 l; 1 mile = 1.609 km.

The well-known increasing curve of gasoline consumption with time is shown in figure 1.1. The figures for 1973 and 1974 show a peak and a decline, the first for over a decade. The decline is continuing into 1975 but only by some 3% over 1974, whereas the demand for other petroleum fuels has notably dropped by much greater amounts. Figure 1.2 shows the retail price of gasoline and how alarmingly this has increased during the last years, particularly of course in 1974. It is interesting to note that UK duty (and VAT) currently amounts to 37½ p of a 75 p/gal retail price (i.e. 50%) and that this has been a steadily growing amount, even if it is currently relatively less than the 70% that it was in 1971. Figure 1.3 shows how the average wage has changed over the same time span, and a combination of figures 1.2 and 1.3 gives figure 1.4, a plot of the time worked by the average working male to earn enough to buy one gallon of gasoline. After declining steadily for many years, this has sharply increased since 1973 and must be having an effect on consumers' gasoline-buying habits. How long this effect will last remains to be seen, but it is important to note that the current figure is still only equivalent to the situation in the late 1960s, and the fraction of running costs attributable to fuel is still not much more than 50%. For the present, however, this effect on fuel costs seems to matter a great deal to the consumer, at least emotionally, and the incentives for improving the consumers' value for money (i.e. miles driven per £) are increasingly great. When longer-term considerations

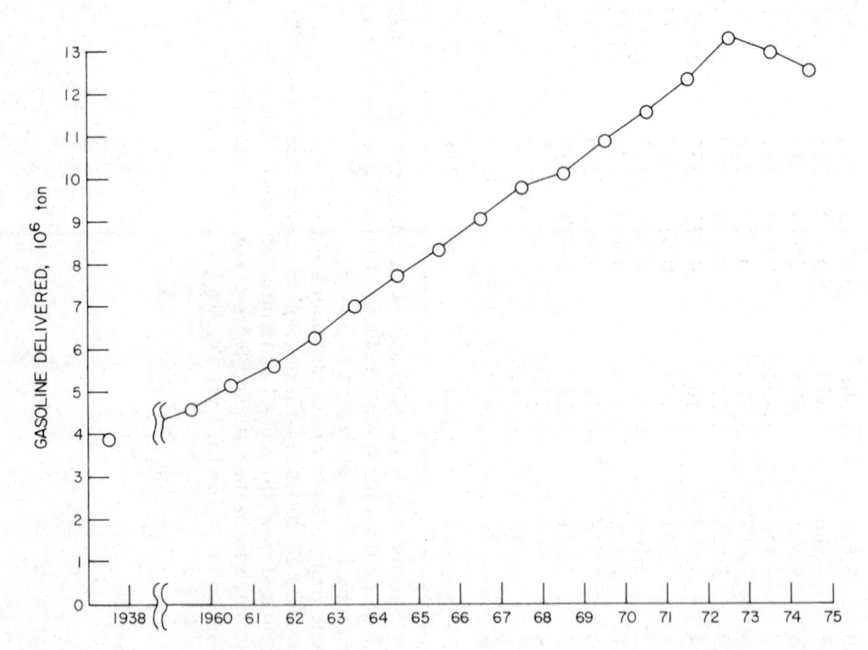

Figure 1.1 Gasoline delivered in GB for cars and motorcycles (1 ton = 1016 kg ≈ 1364 l of gasoline)

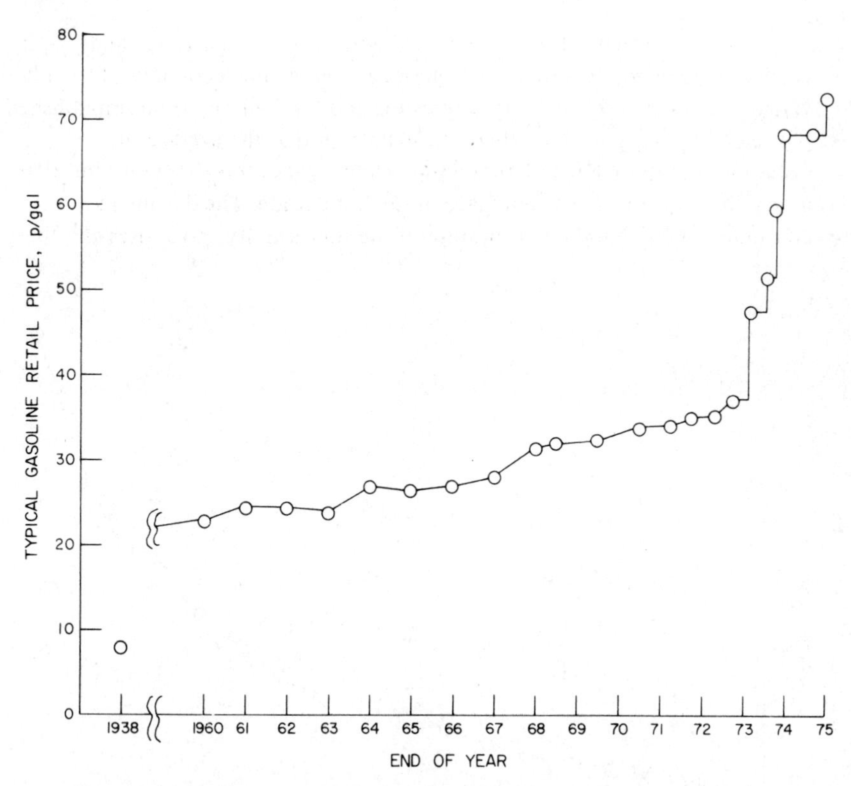

Figure 1.2 Typical retail price for gasoline (premium, inner region) in GB

of possible shortages in energy supply are added to this, the incentive is increased all the more.

However, the picture is not complete with considerations of only gasoline supply and cost: the total miles driven need to be added. Figure 1.5 shows the increasing number of cars in use in the UK (from licence figures for September each year), and the notable flattening of the curve in 1975 is expected since it is known that new car registrations in 1975 declined somewhat. This curve, of course, effectively includes the more widely ranging rates of car production (dependent on motor industry supply and consumer wealth) and scrappage (presumably lower when economic times are harder), and so it is to be expected that the curve will not dip down in concert with the gasoline curve (figure 1.1). However, the miles driven per car per year probably will dip further in 1975: the trend has already started in 1974 (see figure 1.6), giving the first significant decline in average annual distance travelled for many years. Prior to 1974, the trend has been for a steady increase in the distance driven per car, and the curtailment in 1973–4 is probably due to some combination of the limited supply and high price of gasoline.

If the data in figure 1.6 are combined with those in figures 1.2 and 1.5, it is possible to estimate how the UK national average fuel economy of a car has varied over the years. This very interesting and, we believe, so far unpublished plot is shown in figure 1.7, where it can be seen that the average fuel economy is at quite a high level (about 30 mile/gal or 9.4 l/100 km) but that a very small decline has taken place in the last decade. The decline has continued in 1973 and 1974, in spite of the fuel scarcity, price rises and the

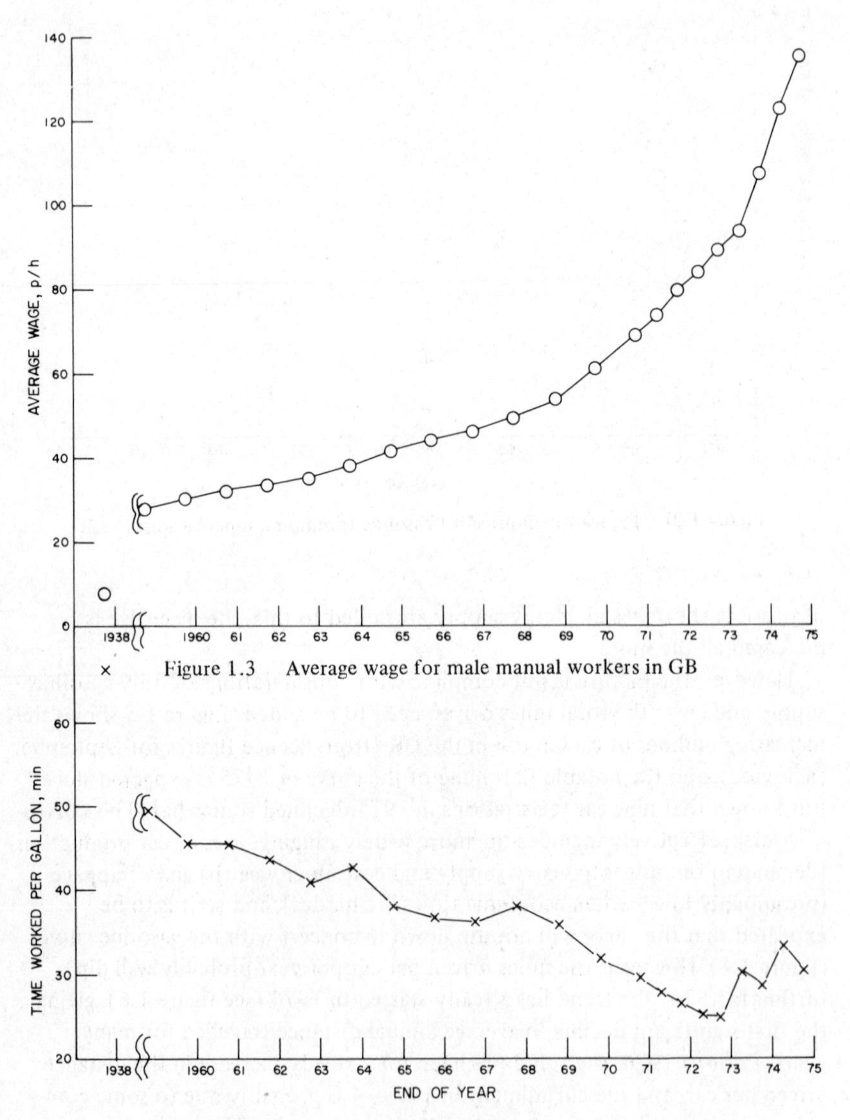

Figure 1.3 Average wage for male manual workers in GB

Figure 1.4 Time worked to buy a gallon of gasoline in GB

publicity given to fuel-saving measures. The factors that influence this national average fuel economy are not yet well defined, though it appears that trends in the model mix and the average distance travelled are both working towards decreasing average fuel economy and that they outweigh any small gains made recently by other means such as speed limits, different driving habits or better maintenance of the vehicle. (It should be noted that the accuracy of the data leading to figure 1.7 may not be good enough to detect small changes reliably, though the trend does appear to be adequately defined. The situation therefore points out the need for improving the accuracy and scope of this rather important measure.)

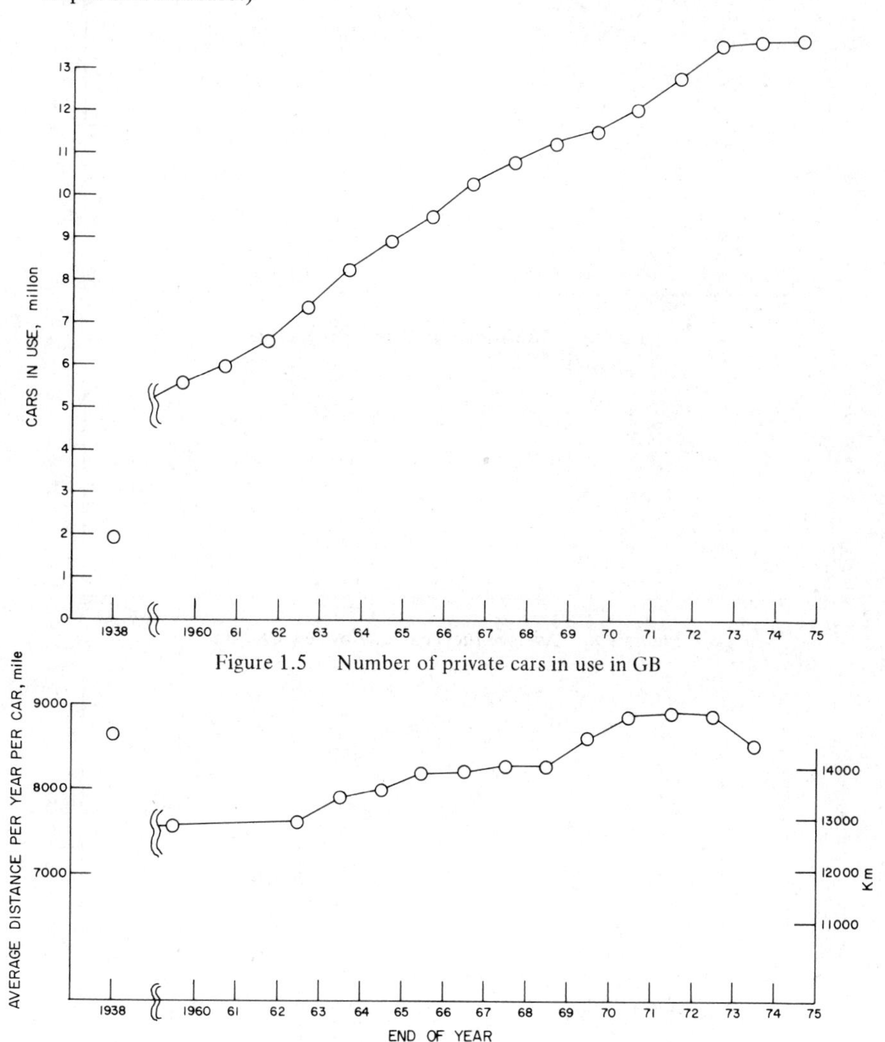

Figure 1.5 Number of private cars in use in GB

Figure 1.6 Average distance per year per car in GB

From all points of view it is desirable that economy shows a significant improvement in the next few years. In the US a start has been made, and figure 1.8[8] shows that for new cars at least the corner has been turned. However, the effects of such improvements will take some time to work through to the curves in figure 1.9, which show the continuing decline up to 1973 in the average economy of cars in use, both as calculated from national figures and also as calculated from the US Environmental Protection Agency's (EPA)

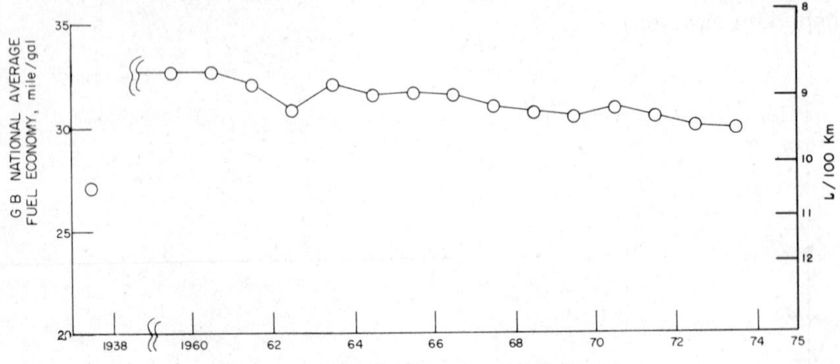

Figure 1.7 GB national average car fuel economy

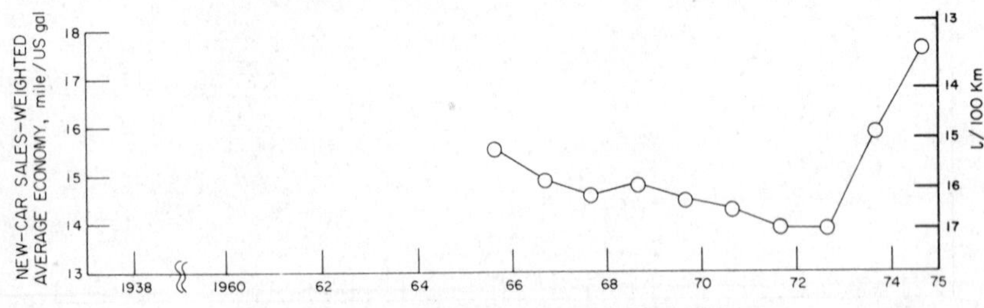

Figure 1.8 Average fuel economy of new US cars

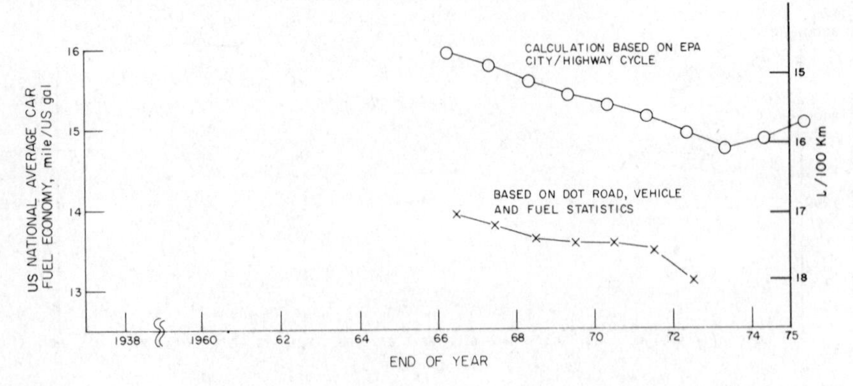

Figure 1.9 US national average car fuel economy

new-car data[8]. Indeed, the data for new cars do indicate that a small gain in national average fuel economy has taken place in 1974 and again 1975. (Incidentally, the fact that the EPA-calculated national average in figure 1.9 gives a line substantially higher than the figure calculated by the US Department of Transportation (DOT) merely indicates that the latest EPA test overestimates the fuel economy on average by this amount. Ironically, their 1975 urban cycle without the highway cycle gave a much better agreement but was vigorously objected to by the US auto industry (see chapter 9).)

How much the UK national average fuel economy will improve from this already fairly high figure over the next few years will be most important. However, for the present the influencing factors need to be more completely understood so that proper decisions concerning the future car population (its size and distribution), car design, taxation (on fuels and vehicles) and fuel supply can be made.

Certainly, it will be very interesting to see how these plots all look in, say, ten years' time.

References

1. *Digest of UK Energy Statistics, 1975*, Department of Energy, HMSO, London (1975)
2. Petroleum Industries Advisory Council annual figures contained in back copies of *Pet. Times*
3. *Highway Statistics, 1973*, Department of the Environment, HMSO, London (1974)
4. J. B. Dunn. Traffic census results for 1972. *Depart. Environ., Crowthorne, Berks., Transport and Road Res. Lab. Rep.*, No. 618 (1974)
5. Figures contained in back copies of *Pet. Times*
6. Department of Employment. Private communication (1975)
   Also partially in *Annual Abstract of Statistics, 1974*, Central Statistical Office, HMSO, London (1974)
7. J. C. Tanner. Forecasts of vehicles and traffic in Great Britain: 1974 version. *Depart. Environ., Crowthorne, Berks., Transport and Road Res. Lab. Rep.*, No. 650 (1974)
8. T. C. Austin, R. B. Michael and G. R. Service. Passenger-car fuel economy trends through 1976. *Soc. Automot. Eng. Pap.*, No. 750957 (1975)

2 Principles Governing Fuel Economy in a Gasoline Engine

D. R. BLACKMORE

2.1 Introduction

The power unit in the majority of passenger cars on the road today is still the gasoline engine. As the name implies, its primary function is to produce the appropriate power as and when the driver commands. However, the demands of the current political and economic climate have brought back into focus yet again the increasing need for this power unit to be an efficient one.

How well this gasoline engine converts fuel energy into mechanical energy is dependent on a large number of practical factors, though these in turn are based on fewer fundamental principles. It is the aim of this chapter to consider these fundamentals in relation to the fuel, the engine and the vehicle.

2.2 Definitions

It will be helpful to begin with the following definitions to clarify what is meant by fuel consumption and how it is measured.

The proper and commonly used basic measure of fuel consumption of an engine at any given constant speed and load condition is the gravimetric specific fuel consumption (sfc). This may be related to either the power (bsfc) as measured on the output of the dynamometer brake or the power (isfc) as measured or computed from the work done (indicated) on the pistons, i.e. ignoring any friction or pumping losses. Thus

$$\text{bsfc} = \frac{M_f}{P_b} = \frac{\text{constant}}{Q_c \eta_b} \tag{2.1}$$

where M_f is the mass flow rate of the gasoline, P_b the brake power, Q_c the gravimetric heat of combustion of the fuel and η_b the brake thermal efficiency (bthe) of the engine process.

While the bsfc is the practical measure of the fuel consumption of an engine, the more fundamental measure isfc is often used:

$$\text{isfc} = \frac{M_f}{P_i} = \frac{\text{constant}}{Q_c \eta_i} \tag{2.2}$$

where P_i is the indicated power and η_i is the indicated thermal efficiency (ithe).

The more popular commercial measure of gasoline consumption is in volumetric terms:

$$\begin{aligned} \text{fuel economy} = \frac{\text{distance}}{\text{volume}} &= \frac{P_b \rho_f K}{M_f} \\ &= K \rho_f Q_c \eta_b \\ &= K Q_{cv} \eta_b \end{aligned} \tag{2.3}$$

where ρ_f is the density of the gasoline, K is a parameter defined by the driving conditions of the car and Q_{cv} is the volumetric heat of combustion of the fuel.

These three expressions show very clearly that fuel consumption is dependent on fuel heating value, on engine efficiency and on vehicle efficiency (governed to a large extent by driving patterns[1]), and so these three areas will now be discussed in turn.

2.3 Fuel Heating Value

For typical gasolines the volumetric heating value varies by a small amount, and this results in a fairly direct relationship with fuel consumption (see chapter 4). Figure 2.1 shows the values of lower (or net) volumetric heating value for a range of typical gasolines[2] plotted against specific gravity (SG). The linear relationship can be described by the empirical relationship

$$Q_{cv} = 11.2 + (28.0 \times \text{SG}) \text{ MJ/l}$$

Thus fuels with higher specific gravities show higher volumetric heating values.

In gravimetric terms, the relationship is inverted, figure 2.1. This has the effect of making engine fuel consumption data, quoted usually in gravimetric bsfc terms, inversely dependent on the fuel specific gravity: it therefore means that care should be taken in converting such engine data to equivalent volumetric (or distance/volume) terms.

Although this effect of the heating value of a gasoline is a most pronounced one and exerts a dominant effect on fuel economies quoted for different fuels under fully warmed-up conditions, it is by no means the only possible or actual fuel effect under all conditions the car experiencies, as will be made clear in later chapters. Effects of fuel volatility and additives are known, and it is possible to conceive that further though smaller effects exist from a variety of other fuel-related features, especially when the time dependence of fuel consumption throughout the life of a vehicle is considered. Closely tied in with the effect of different fuels is the effect of fuel properties on metering by the carburetter, and the fact that, although most carburetters respond similarly to fuel property changes, there are some exceptions, with a consequent different effect on fuel economy.

Although the concept of the heating value of a fuel is straightforward, the measurement of it is rather more difficult. It is inconvenient (and not very precise, unless great care is taken) to use a bomb calorimeter, and so a very useful method has been developed whereby a gas chromatographic analysis of the gasoline is made and the heating value of the gasoline arrived at by summing the literature values[3] of the approximately 200 component hydrocarbons. It is by this procedure that the data plotted in figure 2.1 were obtained.

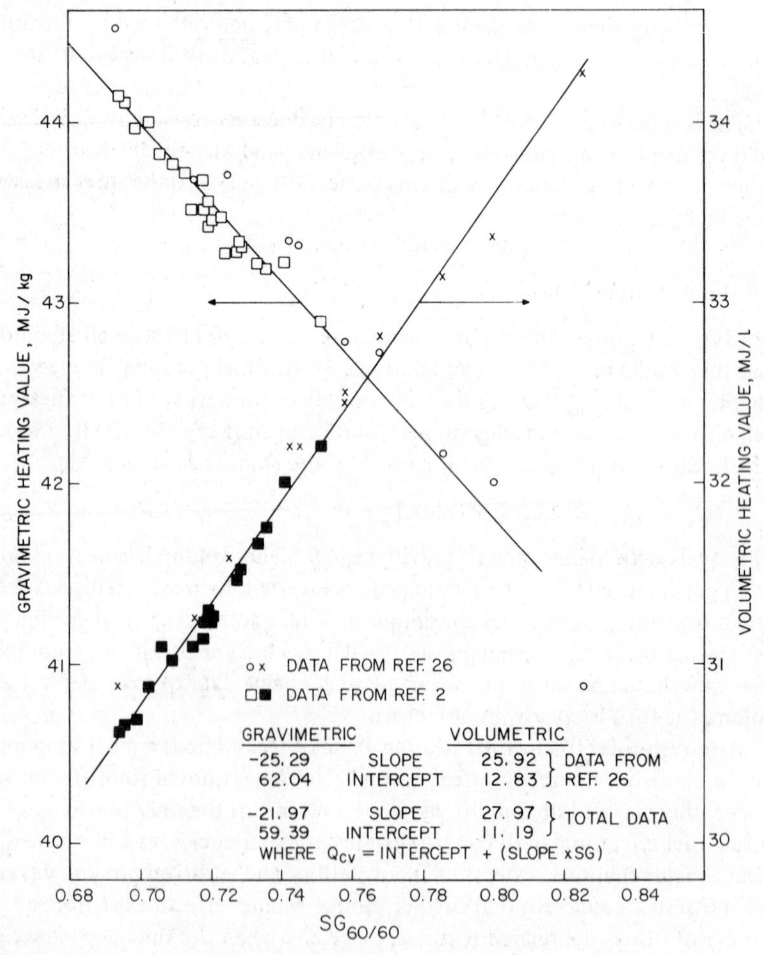

Figure 2.1 Variation of gasoline lower heating value with specific gravity (SG) for full-range gasolines

2.4 Engine Efficiency

In its ideal form the gasoline engine runs on an Otto or constant-volume cycle. Since this cycle in its ideal air version is based only on a few simple assumptions, its thermodynamics may be fairly easily handled mathematically, and a good idea may be gained of its general behaviour, particularly with respect to efficiency and power output*. However, the simpler the assumptions and the resulting analysis, the greater is the danger of mistaken conclusions, and so the cycle analysis has been progressively improved over the years with inclusion of the effects of the fuel and combustion products on the thermodynamics of the process. This subject is treated very clearly in a number of engine textbooks[5–8], in particular Taylor's[5], and we shall only give an outline of the salient features here. Currently, modelling work on the kinetics of the combustion processes occupies workers on this topic, and in particular the emphasis has been towards exhaust emission predictions. Further work is also in progress on such topics as wall quenching, heat loss and the gas dynamics of inlet combustion chamber and exhaust mixture flows.

2.4.1 The Air Cycle

If the working medium is assumed to be a perfect gas having the specific heat and molecular weight of air at room temperature and if it is assumed to go through a cycle in which a given amount of heat is added per unit mass of air at constant volume (figure 2.2), then it can very readily be shown that the indicated efficiency is given by

$$\eta_i = 1 - r^{1-\gamma} \tag{2.4}$$

where r is the compression (or, even better, the expansion) ratio and γ is the specific heat ratio C_p/C_v, which for air at room temperature is 1.40. The efficiency of this cycle is therefore dependent only on r, the compression ratio (figure 2.3) and, it should be emphasized, is independent of the amount of heat added and of the initial pressure, volume and temperature.

The indicated mean effective pressure (imep) of this cycle may also be readily described as

$$\text{imep} = \frac{Q' p_i}{T_i(1 - r^{-1})} \eta_i \tag{2.5}$$

$$= \frac{Q' p_i}{T_i C_v} \frac{1 - r^{1-\gamma}}{(\gamma - 1)(1 - r^{-1})} \tag{2.6}$$

*It should be noted that such an analysis does not take into account the use to which any rejected heat energy in the exhaust or coolant may be put. Generally, this use has been small and has been limited to accessories like the vehicle heating system, but it is possible that this energy could be used to help improve the efficiency of the combustion process. For example, turbocharging by use of exhaust expanders has been recently suggested[4], but retransmitting the saved energy back to an automobile engine poses a very difficult problem. Another example is the use of the exhaust heat to improve fuel–air mixture quality; yet another is to use the heat to reform fuel endothermically to gaseous products that would permit a more efficient engine tuning.

where Q' is the heat added per mass of gas. In contrast with the expression for efficiency, which depends only on r and γ, this expression shows that imep depends directly on Q', r and p_i (the inlet pressure, see figure 2.2) and inversely on C_v, γ and T_i (the inlet temperature). A very significant basic design feature is the ratio (imep/p_3) of imep to peak pressure, since this determines in general the maximum power available for a given peak pressure, and the strength of the engine structure is directly related to the peak pressure it has to withstand. Figure 2.4 shows the decrease of imep/p_3 with increase of compression ratio and indicates how the potential maximum power for a

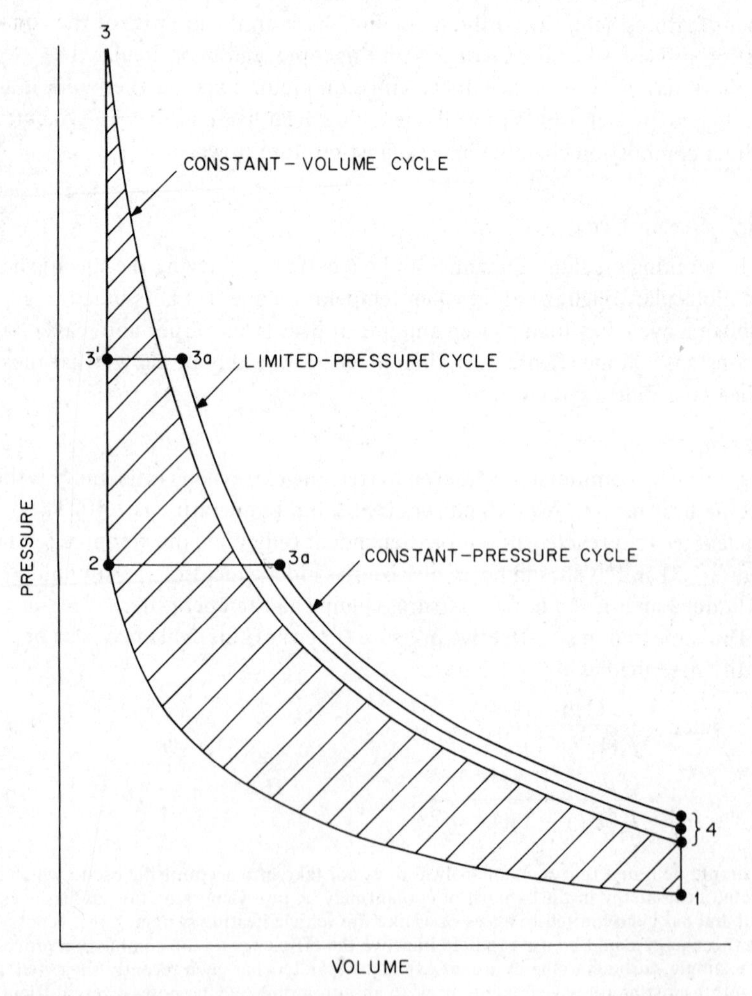

Figure 2.2 Schematic p–V diagrams comparing a constant-volume cycle with a constant-pressure and a limited-pressure cycle

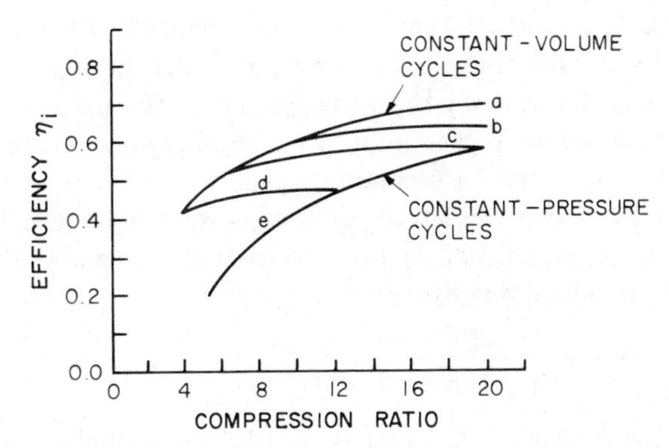

Figure 2.3 Efficiency of constant-volume, limited-pressure and constant-pressure air cycles compared with equal heat input at each compression ratio[5]: curve a, constant-volume cycles; curve b, limited-pressure cycles, $p_3'/p_1 = 100$; curve c, limited-pressure cycles, $p_3'/p_1 = 68$; curve d, limited-pressure cycles, $p_3'/p_1 = 34$; curve e, constant-pressure cycles

given engine structure gets significantly worse as the compression ratio increases. This fact has to be borne in mind when in the practical situation the gains in efficiency with compression ratio are being considered.

One of the chief values of these cycle analyses is that different cycles may be readily compared. Figures 2.3 and 2.4 show this for variations of indicated efficiency and imep/p_3 with compression ratio for the constant-volume Otto cycle and for the constant-pressure cycle, with equal amounts of heat added per cycle. Clearly, for a given compression ratio (say $r = 10$) the

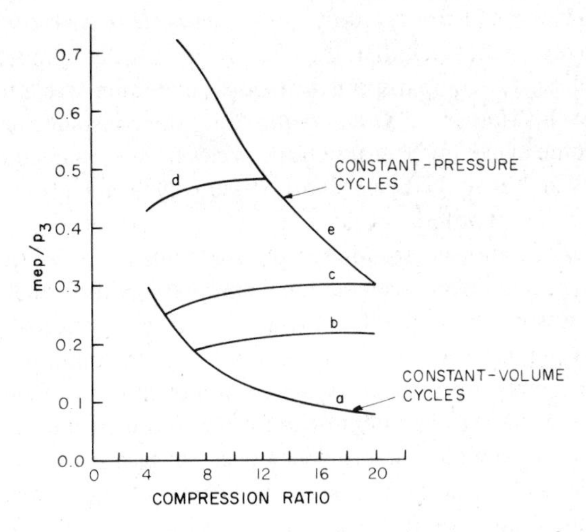

Figure 2.4 Ratio of mean effective pressure to maximum pressure for the cycles of figure 2.3: curves a, b, c, d, e, as in figure 2.3

constant-volume cycle is substantially more efficient (by 40%) than the constant-pressure cycle, though figure 2.4 shows that the constant-pressure cycle possesses a dramatically larger imep/p_3 ratio (by some 400%), i.e. it is capable of much greater power output for a given engine structure.

Because the diesel cycle is best represented by a limited-pressure cycle (i.e. one in which the heat is added partly at constant volume and partly at constant pressure, see figure 2.2), this case has also been analysed. The indicated efficiency is given by

$$\eta_i = 1 - r^{1-\gamma} \frac{\alpha\beta^\gamma - 1}{(\alpha - 1) + \gamma\alpha(\beta - 1)} \tag{2.7}$$

where $\alpha = p_3/p_2$ and $\beta = V_{3a}/V_3$ (see figure 2.2 for a definition of the subscripts). This efficiency is plotted in figure 2.3 for various p_3/p_1 ratios, and, as expected, intermediate curves can be seen as the limited-pressure cycle approaches the constant-volume one. However, it is noteworthy how independent of r the limited-pressure efficiency is at higher compression ratios. Figure 2.4 shows similar intermediate curves for the imep/p_3 relationship, where imep/p_3 now slowly increases with compression ratio, indicating a small potential power gain with increasing compression ratio for this engine configuration.

These comparisons are all based on the assumption of equal quantities of heat being added at equal compression ratios. If the equations are turned round and if the calculations are made on the basis of equal quantities of heat being added at equal maximum pressures (i.e. roughly equivalent engine structures), then the constant-volume cycle is restricted to a much lower compression ratio and also now to a lower efficiency than that of the constant-pressure cycle. A similar comparison is found for the case of equal maximum pressure and work output: the diesel cycle of high compression ratio has a greater efficiency than the Otto cycle of lower compression ratio. Yet another basis for comparison is with equal maximum cycle temperature and pressure: this time more heat is required by the constant-pressure than the constant-volume cycle, but, because both cycles have the same heat rejection, the efficiency and imep of the constant-pressure cycle are again greater than those of the constant-volume cycle.

All three of the above bases for comparison come nearer to the practical situation than does a comparison at equal compression ratio and therefore indicate one reason why in practice the diesel engine is preferred to the gasoline engine when improved economy is important. In common usage, though, the diesel is normally operated at higher compression ratios (from 14:1 to 20:1) and also at higher maximum pressure (therefore engine strength), thereby gaining a little more in efficiency over the gasoline engine at the expense of extra manufacturing cost of the engine. (A more complete appraisal of the fundamental reasons why the diesel is usually a more efficient engine in practice is given in appendix E.)

2.4.2 The Fuel–Air Cycle

One of the principal limitations of the air cycle simulation is that the thermodynamics of the working fluid are grossly oversimplified, in particular in relation to its specific heat; in reality this is strongly dependent on temperature (figure 2.5), and so therefore is γ (i.e. C_p/C_v) in equation 2.4. Figure 2.6 shows that, when the variable specific heat of the air is introduced into the model, the indicated efficiency drops from 0.570 to 0.494 for a compression ratio of 8:1.

However, to carry the simulation yet further it is necessary to take into account the thermodynamic contribution that the fuel and its products make to the working fluid at all parts of the cycle. The fuel–air cycle is therefore constructed as follows.

(i) On the compression stroke (1 → 2 in figure 2.2) the unburnt fuel is mixed with air, water vapour in the air and residual gas and is compressed adiabatically to point 2. This involves knowing the composition and thermodynamic properties of the fuel (specific heat, latent heat of vaporization) as well as those of air, residuals and water vapour.

(ii) The mixture is combusted at constant volume (2 → 3 in figure 2.2); the heat of combustion at constant volume, the yield of product molecules per molecule of combustible mixture and the specific heat of these products are required.

(iii) The hot equilibrated products of combustion are allowed to expand adiabatically (3 → 4 in figure 2.2), and the work of expansion is calculated.

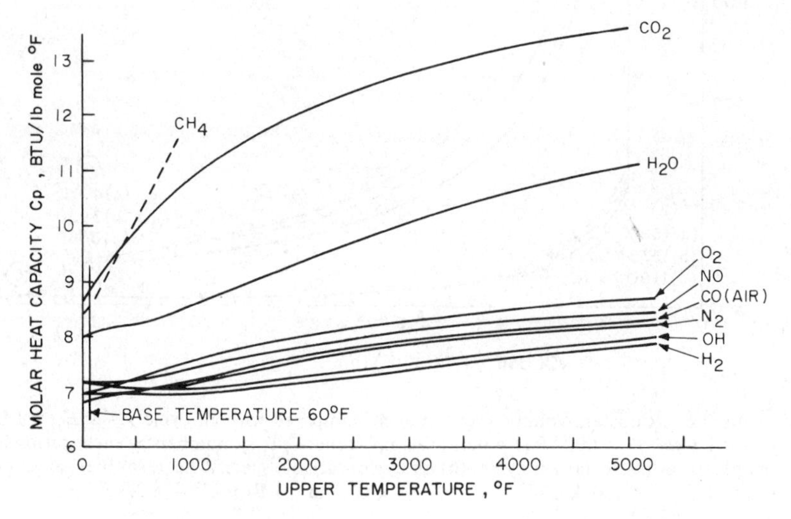

Figure 2.5 Molar heat capacity at constant pressure of gases above 60 °F, quoted as averages between 520 °R and abscissa temperature[5] (1 Btu/lb mole °F = 1.292 kJ/mole K; 1 °F or 1 °R = 0.555 K)

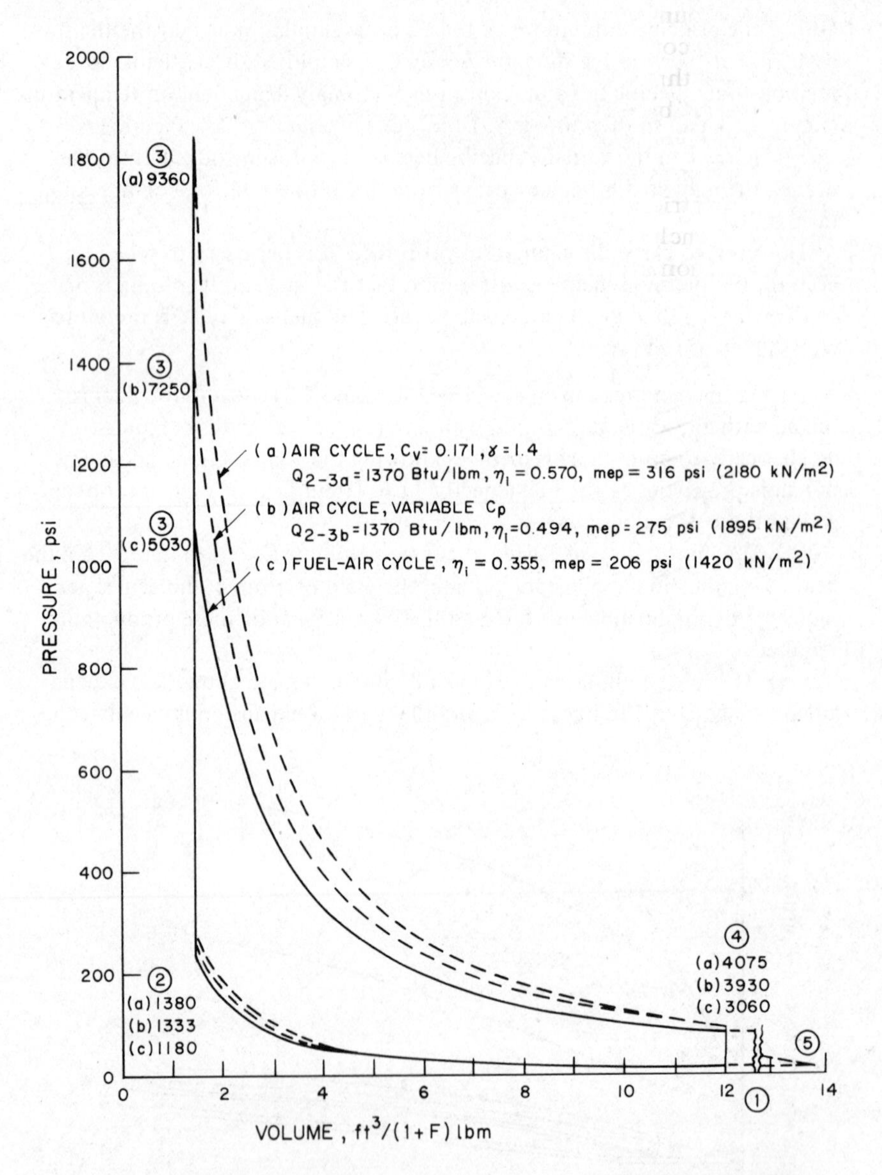

Figure 2.6 Constant-volume fuel–air cycle compared with air cycle. $r = 8$; $F_R = 1.2$; $p_1 = 14.7$ psi, $T_1 = 600\,^\circ$R, $f = 0.05$; fuel is octene, C_8H_{16}; numbers at each station are temperatures in °R. (a) Air cycle; (b) air cycle, variable specific heat; (c) fuel–air cycle (1 psi (lb/in²) = 6.89 kN/m²; 1 ft³ = 2.83 x 10⁻² m³; 1 Btu/lb = 478 J/kg; 1 °R = 0.555 K)

It should be noted that this treatment is essentially a thermodynamic one, taking no account of finite rates of reaction nor of spatial effects. It assumes instantaneous combustion at top centre and assumes no heat losses to or from the wall throughout the cycle.

In Taylor's[5] book, he describes this simulation procedure in detail and provides four working charts with which the calculations can be carried out for any fuel/air ratio F or mixture strength ($F_R = F/F_c$), where F_c is the fuel/air stoichiometric ratio for any hydrocarbon fuel. Figure 2.6 shows the marked effect of the inclusion of these fuel-related factors on the indicated efficiency: at a compression ratio of 8:1 the efficiency drops dramatically from 0.494 for

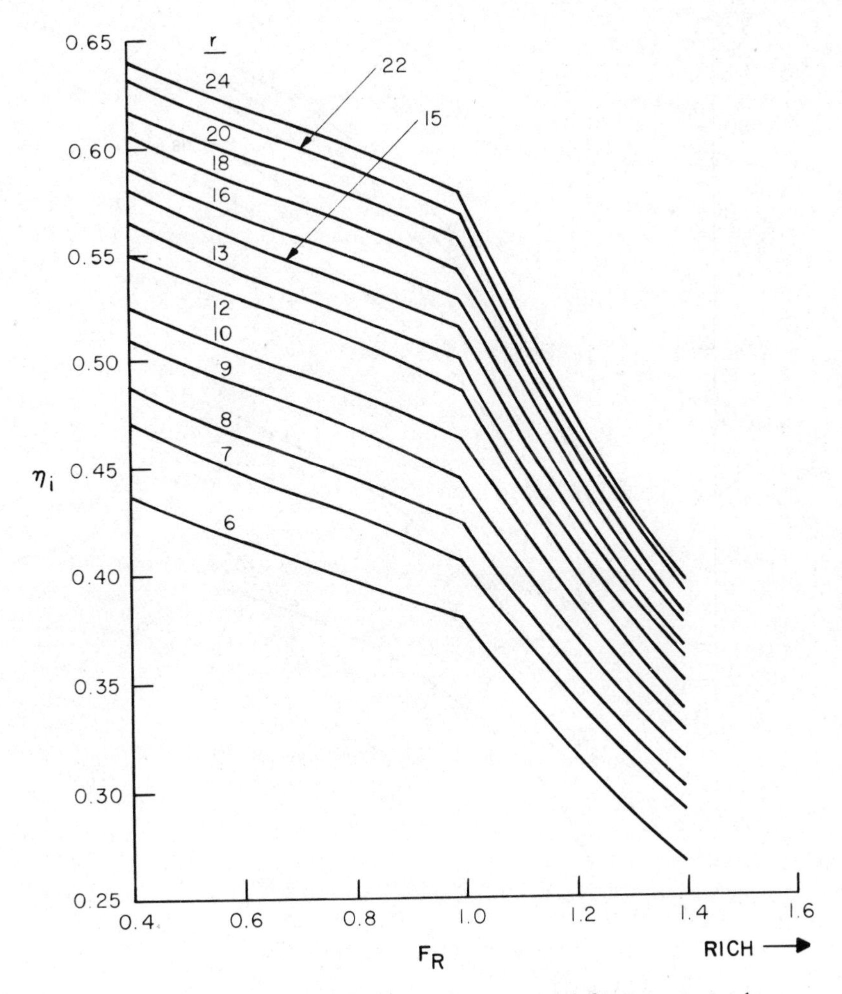

Figure 2.7 Variation of efficiency with mixture strength for a constant-volume fuel–air cycle with 1–octene fuel[9]

the air cycle with variable specific heat to 0.355 for a rich (F_R = 1.2) mixture of octene and air.

The very strong effects of both mixture strength and compression ratio on indicated efficiency are well illustrated in the very important figures 2.7 and 2.8. It is clear that for all values of compression ratio the efficiency continues to increase as the mixture leans off, and this may be attributed to the progressive approach of the fuel–air cycle to the air cycle as the proportion of fuel

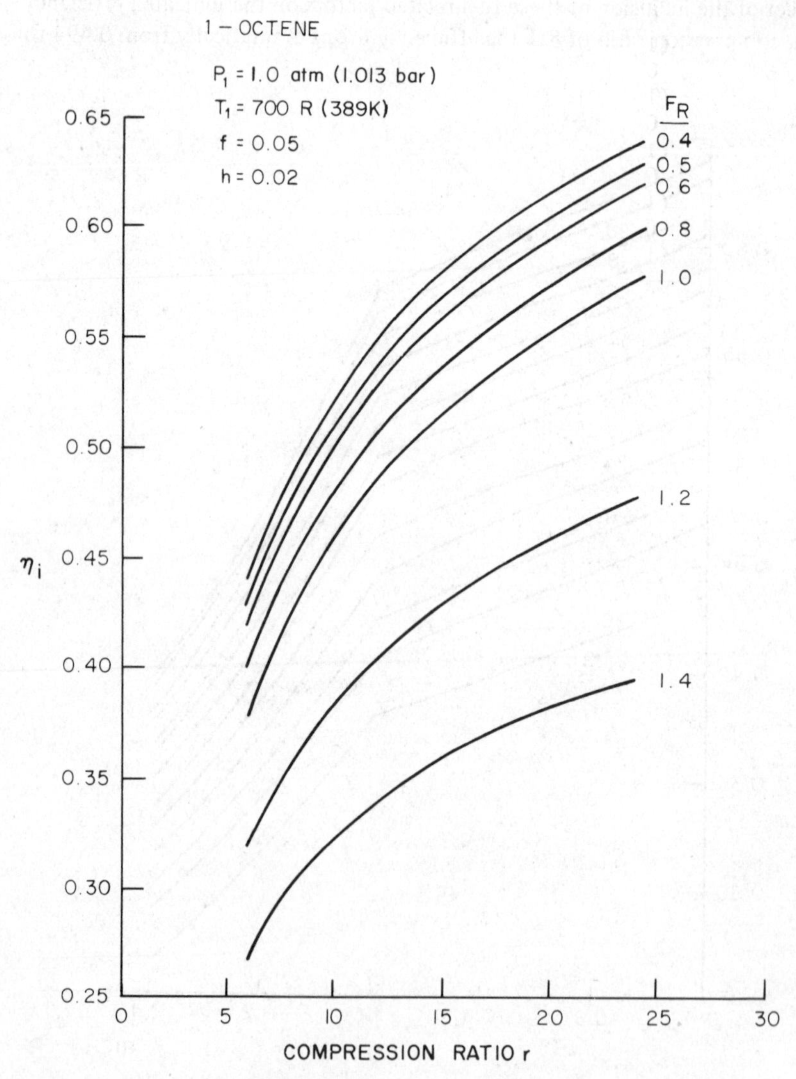

Figure 2.8 Variation of efficiency with compression ratio for a constant-volume fuel–air cycle with 1–octene fuel[9]

Table 2.1 Effect of different compression ratios and fuel/air ratios on thermal efficiency (calculated for theoretical air cycle and fuel–air cycle) (from Edson and Taylor[9])

Compression ratio	Air cycle efficiency η_i	Fuel–air cycle efficiency at F_R =				
		0.6	0.8	1.0	1.2	1.4
7.0	0.540 (95.6)	0.448	0.428	0.403 (95.3)	0.340	0.287
8.0	0.565 (100)	0.470 (111.1)	0.449 (106.1)	0.423 (100)	0.357 (84.4)	0.300 (70.9)
9.0	0.585 (103.5)	0.488	0.467	0.442 (104.5)	0.372	0.312
10.0	0.602 (106.5)	0.503	0.483	0.459 (108.5)	0.385	0.322
11.0	0.617 (109.2)	0.517	0.496	0.473 (111.8)	0.396	0.330
12.0	0.630 (111.5)	0.530	0.508	0.486 (114.9)	0.406	0.338

The bracketed numbers are relative to a compression ratio of 8.0 and at a stoichiometric mixture strength in the case of fuel–air cycles.

diminishes to zero. It should be noted that although the efficiency–compression ratio relationship appears to become more pronounced as the mixture gets leaner, in proportional terms it is in fact virtually constant.

Table 2.1 sets out the quantitative data of figures 2.7 and 2.8 for easy reference. Included are the relative changes from a stoichiometric mixture at a compression ratio of 8:1, which show the relative magnitude of theoretical gains to be had from changes in compression ratio or mixture strength.

2.4.3 *Simulated Engine Cycles*

The advent of the computer has brought about a significant advance in the science of cycle simulation. In particular, the inclusion of time-dependent phenomena such as flame propagation rates and heat transfer effects has been attempted, though not yet satisfactorily, and the relation of the whole to engine variables such as speed, piston motion, combustion chamber shape, spark plug location, spark advance, engine displacement, stroke/bore ratio, mechanical friction, pumping losses, etc., has been explored. The result of much of this work has been to arrive at quantitatively better engine predictions of efficiency, imep, emissions, etc., but has not altered the directional trends of efficiency, compression ratio and mixture strength mentioned above.

2.4.4 *Reasons for Discrepancies between an Actual Cycle and an Equivalent Fuel–Air Cycle*

The theoretical yardstick against which comparison of an actual cycle may be made most conveniently is the equivalent fuel–air cycle[5]. This is computed as for the fuel–air cycle but after equivalence with an actual cycle has been established by making the charge compositions and densities as similar as possible in the two cycles.

Figure 2.9 shows some schematic indicator diagrams of actual cycles compared with their equivalent fuel–air cycles and illustrates the losses that can occur.

(a) Exhaust losses occur because the exhaust valve is designed to open a little before bottom centre to help in the exhaust scavenging. Taylor[5] estimates these losses as 2%.

(b) 'Finite piston speed' losses occur because the combustion process necessarily takes time, during which the piston moves up to and away from top centre with the result that the p–V diagram is somewhat smoothed out. Taylor[5] estimates these losses at about 6%. He distinguishes them from the negligible losses that can be attributed to progressive burning alone, in which the slightly enhanced work done by the first-burnt portion, with its subsequent recompression and expansion path, is averaged with the slightly decreased work done by the last-burnt portion.

(c) Heat losses that affect the efficiency of the cycle occur, as Ricardo and Hempson[8] are at pains to point out, only in the compression and expansion cycles, and even then they are most damaging if they occur near top centre. Inspection of the p–V indicated diagram provides one way of quantitatively

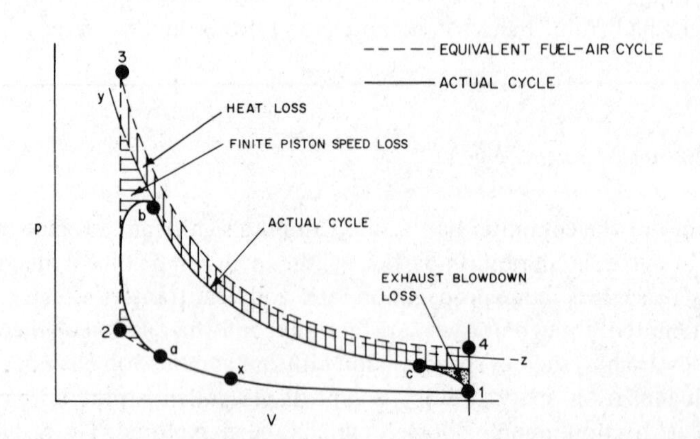

Figure 2.9 Losses in an actual cycle compared with the calculated equivalent fuel–air cycle (showing the apportioning between heat loss and 'finite piston speed' loss)[5]: y–z is isentropic through point b; a–b is the non-ideality caused by finite piston speed; b–c is the non-ideality caused by heat loss during expansion; c–1 is the non-ideality caused by exhaust blowdown loss.

assessing the magnitude of these losses. Taylor[5] gives a value of 12% for the loss in indicated efficiency in one sample and a nearly zero value in another (air-cooled engine) example. An estimate by Ricardo and Hempson[8] also arrives at 12%, though they point out that even if these losses were recovered by some means the higher cycle temperatures would lead to increased specific heat and dissociation losses that would reduce this efficiency gain to around 7.5%. They conclude that such heat losses play a relatively small part in the engine cycle and that coolant heat balance relates more to heat lost in the exhaust process.

(d) Other possible losses have been considered and reckoned to be negligible in a properly operating engine. Leakage losses can be caused by blowby past the piston or even past the inlet or outlet valves. Losses due to incomplete combustion, caused by failure to burn all the charge by the time the exhaust valve opens, are small, as seen from the low exhaust emissions of unburnt hydrocarbons in modern engines (though it has to be said that the quench layer from which these emissions come may be a significantly large fraction of the fuel in the critical part of the cycle near top centre). Frictional heat generation by the pistons is a complicated factor, for properly such losses do not contribute to the indicated thermal efficiency but rather to the brake thermal efficiency. However, the heat involved (up to 1.5% of the total heat of the fuel according to Ricardo and Hempson[8] or say 4.5% of the indicated efficiency), which flows at least in part to the cylinder walls, will to a secondary but very minor extent affect the wall temperature and main heat loss.

2.4.5 *Further Reasons for Discrepancies between an Actual Cycle and an Equivalent Air Cycle*

Often the efficiency of an actual cycle is compared with an equivalent air cycle, since the latter is so readily calculable. Although this is not the most valid comparison in view of the fairly ready availability of the equivalent fuel–air cycle, yet a study of the discrepancies does serve to show up some of the fundamental reasons for the behaviour of fuel–engine systems.

(a) Specific heat increases with temperature (see figure 2.5), and this is most pronounced with water vapour. This indicates the desirability of working at the lowest possible maximum temperature.

(b) Chemical dissociation of molecules at high temperatures with the consequent very large absorption of heat (and therefore reduction in temperature and pressure) is another cause of loss of ideality. Ricardo and Hempson[8] point out that, although the effect is not large in terms of loss of efficiency (unless there is a failure of the atoms and radicals to recombine early in the expansion stroke, as appears to be the case to some extent), yet the effect is there and is more pronounced with CO_2 than with H_2O. The effect therefore counterbalances the specific heat disadvantages of H_2O relative to CO_2, and to a large extent this is the reason why thermal efficiency is so little affected by fuel composition.

(c) Product mole yield for real fuel–air systems is not unity but slightly greater, and it increases as mixtures get richer. This has the effect of giving a slightly improved efficiency and power output for real systems. There is very little difference between the product mole yields for most normal hydrocarbon fuels, and there is only a small effect depending on whether the fuel enters the combustion chamber as liquid or as vapour.

Since the efficiency of the equivalent air cycle is given by

$$\eta_i = 1 - r^{1-\gamma} \tag{2.4}$$

and the only variable associated with the combustion process is γ, the ratio of the specific heats, it is often found convenient to describe the efficiency of actual cycles by the above equation but with a modified value for γ (such as 1.21 instead of 1.396, which is the value for air at room temperature). Alternatively, attempts have been made to describe the processes by using one γ value for compression of the fuel–air mixture (1.33) and another for expansion of the products (1.15–1.25).

While this approach is useful for its qualitative convenience, its weakness as a quantitative prediction is well exemplified in the claims made for the effect of exhaust gas recirculation (EGR). It has been suggested[10] that the main effect of EGR is to lower the maximum cycle temperature and thereby to give higher effective values of γ and therefore improved efficiency, even though EGR gases will have a higher specific heat (and therefore a lower γ value) than air or a fuel–air inlet mixture and much the same γ value as the burnt working fluid during expansion.

Calculations can be made on fuel–air cycles. Taylor[5] shows that the increase of exhaust residuals* from a value of from 5 to 10% results in only small temperature reductions and a decrease in η_i only of from 0.355 to 0.348. Clearly the real situation is controlled by the temperature of the recycled gases plus other factors such as decreases in flame speed and extra heat losses incurred, and only much more careful modelling will show which of all these conflicting influences really dominates.

One feature that the use of air cycles brings out is the fundamental gain that results from increasing the compression ratio, though clearly from figures 2.7 and 2.8 it is subject to the law of diminishing returns. A well-known study by Caris and Nelson[11] showed that 17:1 was the compression ratio for peak efficiency, always provided that suitable fuel could be made available (which it cannot (see chapter 4)). This limit appeared not to be related to any increase in friction but rather to arise because of the result of lower combustion rates;

*Residuals are equivalent to recycled exhaust gas, except that their temperature is fixed at a somewhat high value around 2100 °R (1170 K) by the cycle processes.

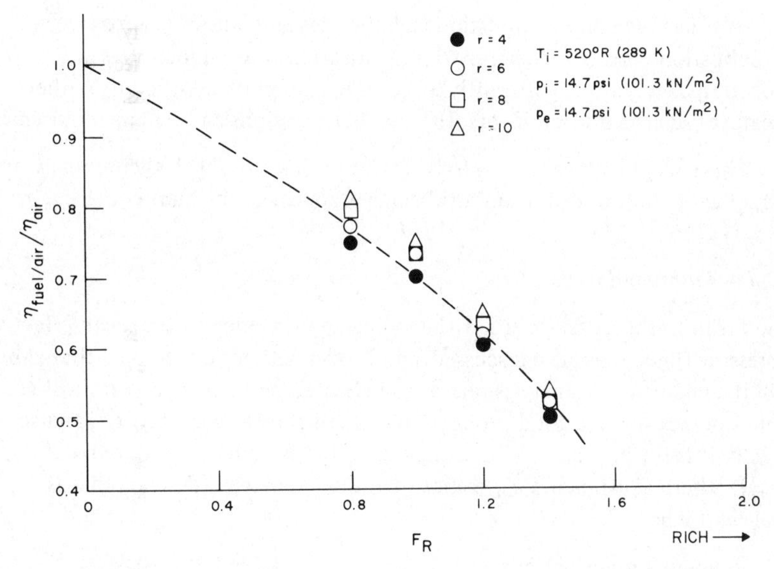

Figure 2.10 Variation with mixture strength of efficiency of fuel–air cycle relative to that of air cycle showing the gain in efficiency as the mixture weakens, independent of compression ratio[5]

it was speculated that the reason for this was connected with the increased surface volume ratio, which can give rise to heat losses, quenching of the flame, etc.

Another feature that the use of air cycles brings out is the fundamental gain that must result as a fuel–air mixture is progressively leaned out, until ultimately the efficiency approaches that of the air cycle itself (figure 2.10). Just how far it is practically possible to go along this path is a question for the engine and fuel designers, but significant advances have been made in the last several years and currently the 'lean-burn' approach is receiving a great deal of attention. As the factors behind the capacity for an engine to burn lean are being studied (see chapter 6), so the following requirements emerge.

(i) The importance of good spatial and temporal mixture homogeneity (both of carburetted mixture and also of this mixture with the exhaust residuals).

(ii) The need for controlled mixture turbulence, for, although the very existence of the high-speed gasoline engine depends on the accelerating effect on the flame of the increasing turbulence generated by the high speeds, it is possible that too great a level of turbulence can affect the ability of the spark to ignite and thereby can limit the extent to which the engine may be leaned out.

(iii) The fact that improved ignition systems (such as higher-energy or multi-spark systems) can aid the ability of mixtures to burn.

(iv) The fact that combustion chamber design features (such as combustion chamber shape, spark plug location, valve dimensions and locations, etc.) also significantly affect the lean limit and of course other features such as quench layers, HC and NO emissions and octane requirement.

There would appear much to be learnt in this area, both theoretically and practically, before real 'lean-burn' engines appear in the market-place.

2.4.6 Pumping Losses

So far in this discussion, either the indicated efficiency or the mean effective pressure (imep) has been discussed, and it is sometimes overlooked that almost all the engine cycle analysis relates to indicated performance. Yet most gasoline engines spend a good proportion of their time at part throttle, and so really it is the efficiency under this mode that is important; moreover it is the brake thermal efficiency η_b rather than the indicated efficiency that is relevant when

$$\text{imep} = \text{bmep} + \text{fmep}$$

and

$$\text{fmep} = \text{pmep} + \text{ramep}$$

where fmep is the friction mean effective pressure, made up of a contribution from pumping mean effective pressure (pmep) and rubbing and accessory mean effective pressure (ramep).

Pumping loss has its basis in the work required to pump inlet gas from inlet pressure p_i to exhaust pressure p_e. The actual p–V diagram as recorded[12] during the suction and exhaust strokes gives a measure of it (figure 2.11). It has two main components, one due to the work required to overcome inlet and exhaust system throttling losses ($p_i - p_e$ factor) and the other due to the work required to move the gases past the valves through the combustion chamber (valve loss factor). There is a third blowby component due to losses on compression and expansion, but this is usually too small to matter.

Figure 2.12 shows[13] the variation in pmep as a function of imep. It is very clear that the total pumping loss increases very strongly as imep decreases (by throttling of the engine) until pmep approaches a value of 15 psi, the maximum that $p_e - p_i$ can possibly reach. Indeed, in its simplest form the 'net' efficiency η_n (i.e. indicated efficiency less pumping losses due to throttling only) is given by

$$\eta_n = \eta_i \left(1 + \frac{p_i - p_e}{\text{imep}}\right)$$

However, in the real case there is an added contribution from valve losses, and so a plot of pumping loss against imep shows that the magnitude of the total

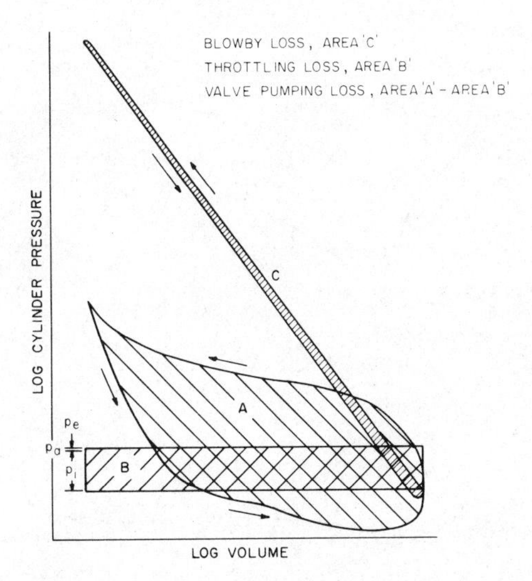

Figure 2.11 Typical single-cylinder pumping diagram under motoring conditions[12]

pumping loss varies from 3.5% at wide-open throttle (WOT) to nearly 100% for a fully throttled idling engine. Figure 2.13 shows[13] how this total pumping loss affects the indicated thermal efficiency for a compact US 10:1 compression ratio car cruising at 40 mile/h (64.4 km/h): the efficiency decreases by 5.5% absolute or about 16% relative. Because the valve losses are present particularly at WOT it is never possible to regain the losses completely;

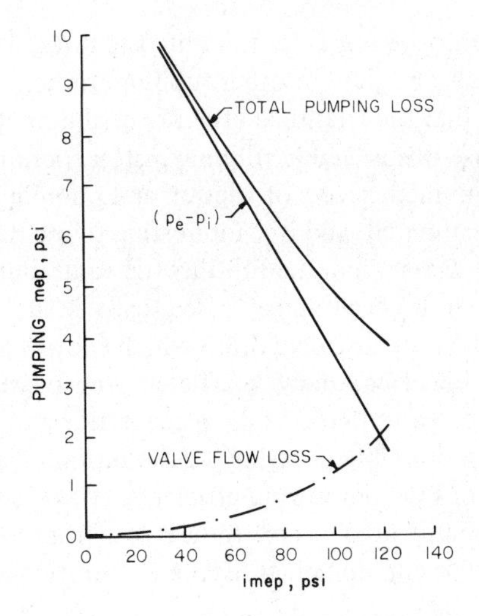

Figure 2.12 Variation of pmep with load, showing throttle valve as the main source of pumping losses at part throttle[13]. Data are for 7:1 compression ratio engine operating at 1600 rev/min (1 psi (lb/in^2) = 6.89 kN/m^2)

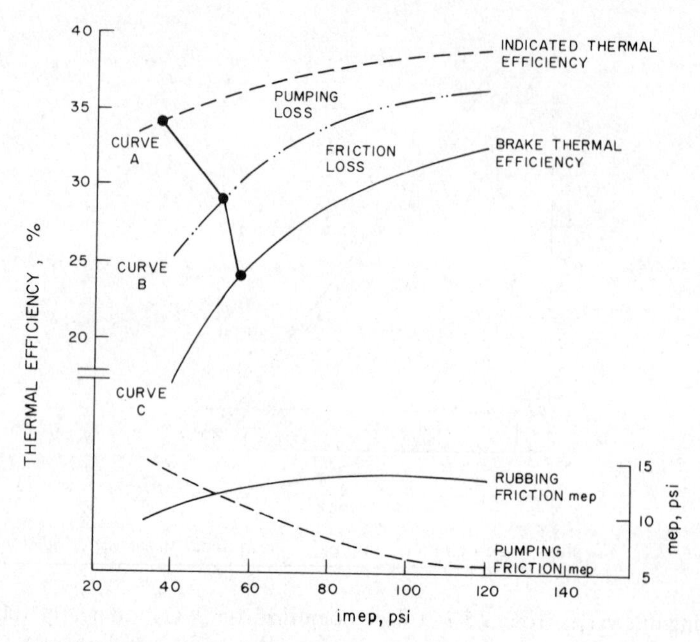

Figure 2.13 Variation of efficiency and mean effective pressure with load (showing relative effects of pumping losses and friction losses)[13]. Data are for a 10:1 compression ratio engine operating at 1800 rev/min (1 psi (lb/in^2) = 6.89 kN/m^2)

2% absolute is the total pumping loss retrievable under these conditions (figure 2.13).

The bsfc at part throttle is greater than the isfc not only because of the pumping losses but also because of other rubbing friction losses. Cleveland and Bishop[13] show that the frictional effect contributes about a 5% absolute loss in η_i, and because this is largely rubbing piston friction it cannot be recovered. If the combined effects of friction and pumping losses are allowed for, the isfc can be estimated, and it is interesting to see that this part-throttle isfc is not very much worse than the full-throttle value and also is fairly consistent over the whole speed range[14] (see figure 2.14). This is in broad agreement with fuel–air cycle calculations which show that part-throttle operation has negligible effect on cycle efficiency or maxium cycle temperature[5]. The increases that do take place in this part-throttle isfc at low loads are probably attributable to residual gas dilution of the charge, which brings about reductions in combustion efficiency (flame speed); at high loads the increases are probably attributable to heat loss from the extra gas turbulence and mixture enrichment from the carburation aimed at maximizing the power.

Table 2.2 attempts to set out the quantitative effect of pumping losses for easy reference. It shows that the simple calculation based on only the factor $p_i - p_e$ gives a fair estimate. It also shows the progressively increasing

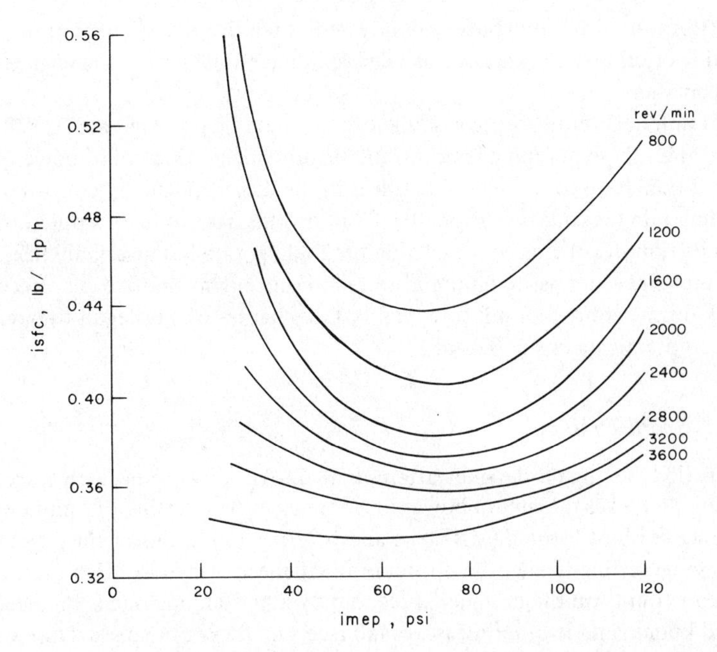

Figure 2.14 Variation of part-throttle isfc with imep at different engine speeds, showing relative constancy at high speeds[14]. At low speeds losses occur both at low loads (heat loss) and at high loads (friction loss) (1 lb/hp h = 0.169 kg/MJ; 1 psi (lb/in^2) = 6.89 kN/m^2)

Table 2.2 Variation of thermal efficiency with load at constant speed (showing quantitatively the effect of pumping losses)

imep, % of maximum	Ideal case (case 1)	Case in figure 2.12 (case 2)	bmep, % of maximum	Case in figure 2.13 (case 3)	
	Relative net efficiency η_n/η_i	Relative net efficiency $\eta_n/\eta_n(1.0)$		Relative brake efficiency $\eta_b/\eta_b(1.0)$	Relative net efficiency $\eta_n/\eta_n(1.0)$
100	1.00	1.00	100	1.00	1.00
80	0.96	0.98	76	0.95	0.98
60	0.90	0.94	52	0.85	0.91
40	0.78	0.86	27	0.65	0.78
20	0.30	–	–	–	–
0	0	–	–	–	–

Net efficiency is {imep – pmep)/imep} η_i, allowing for pumping losses only.
Brake efficiency is {(imep – pmep – ramep/imep)} η_i, i.e. allowing for rubbing friction and accessory losses.
Case 1: ideal, i.e. $\eta_n = \eta_i\{1 + (p_i - p_e)/\text{imep}\}$, maximum imep = 100 psi (689 kN/m^2).
Case 2: $\eta_n = \eta_i(1 - \text{pmep}/\text{imep})$, where pmep is read off figure 2.12.
Case 3: $\eta_n = \eta_i\{1 - \text{pmep} + \text{ramep})/\text{imep}\}$, where the efficiencies are read off figure 2.13.

contribution of pumping losses (expressed as relative net efficiency) to the total friction losses (expressed as relative brake efficiency) as the engine load is decreased.

In practical terms, if the gasoline engine could be operated at WOT all the time, then these pumping losses would be minimized. Control of power output would then have to be achieved, as it is in the diesel engine, by control of fuelling rate (i.e. mixture strength). Hitherto this has not been a practical proposition, for the effect of cutting the fuelling rate has invariably been to run into the lean misfire limit and to render the engine undriveable. Recent work on the control of mixture quality (see chapter 6) has begun to open this area to practical consideration.

2.4.7 Spark Timing

Normally the engine designer attempts to provide an engine with a spark timing that gives maximum power or, more exactly, a setting of 'minimum advance for best torque' (MBT) for all operating conditions of the engine. This same setting is then the optimum for fuel economy also. However, there are constraints on the designer's freedom. A long-standing one is the need to avoid knock, and if other measures fail a certain degree of spark retard will normally suffice to eliminate it. A more recent constraint is the need to control exhaust emissions, and both HC and NO emissions are reduced by spark retard. Consequently, in some modern engines a small but significant amount of spark retard is deliberately introduced, and this has been shown in some cases to result in a measurable fuel consumption penalty. In three 1974 US cars[15], a fuel economy gain of 8.4% resulted from a 10° crank angle (CA) spark advance. A recent calculation[16] in terms of indicated efficiency showed only a 1.6% relative loss for 10° CA spark retard from optimum, and Taylor[5] gives an example where for a single-cylinder engine a 10% relative loss in η_i occurs for about a 20° CA spark retard.

It would thus appear that in the practical case losses do result from the use of retarded timing, and for some reason these appear larger in vehicles than might be expected on the basis of theoretical or laboratory engine work.

2.4.8 Engine Mapping

In the preceding sections we have discussed a number of features of an engine that affect efficiency and have seen that the dominant influences are compression ratio, mixture strength and throttle position, with a lesser effect from spark timing.

The engine builder, having taken due note of these features and doubtless very many more relating to all the other necessary aspects of engine performance (power output, startability, driveability, knock resistance, exhaust emissions, etc.), defines an engine at least in the steady state in terms of

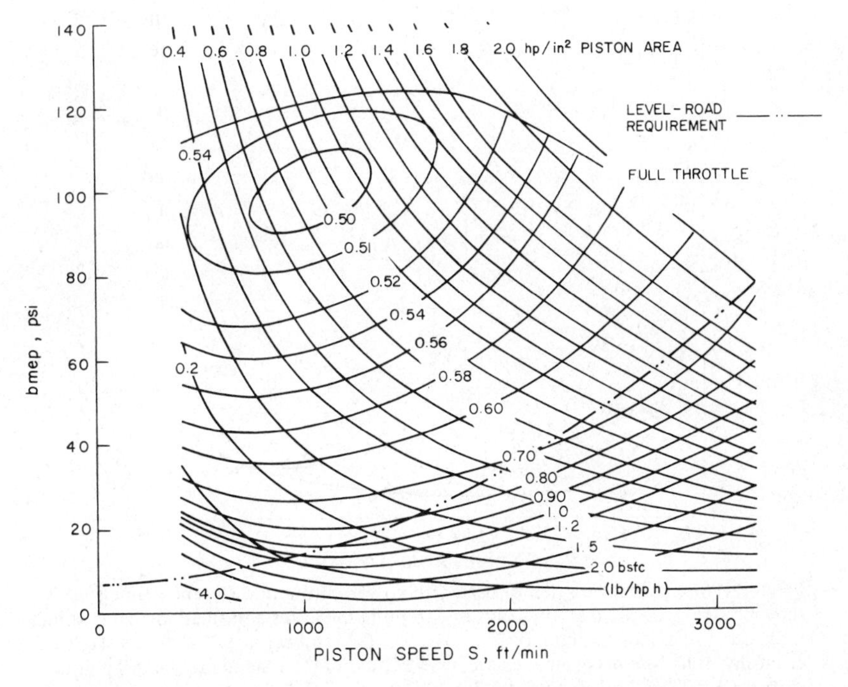

Figure 2.15 Performance map for a typical US passenger-car engine as installed[5] (1 psi (lb/in^2) = 6.89 kN/m^2 : 1 ft/min = 5.08 mm/s; 1 lb/hp h = 0.169 kg/MJ; 1 hp/in^2 = 1156 kW/m^2)

performance maps, a popular one being a plot of bmep against engine speed with contours of lines of equal bsfc (figure 2.15). It should be emphasized that all the tuning of the engine has been completed and that this diagram contains variations in air/fuel ratio, spark timing, etc., according to the designer's choice. Therefore, no two engines have the same map: indeed it is probably true to say that no two supposedly identical engines will have quite the same map. Figure 2.16 gives the same data for a tuned engine[17], but this time it is plotted on different axes (i.e. bsfc plotted against bhp). This method of plotting serves to emphasize the loss in bsfc with light loads.

2.5 Vehicle Efficiency

The next phase of building a road vehicle involves the matching of an engine to the needs of the vehicle via the transmission, and, if good economy is to be a significant objective, then by suitable choice of gearbox, rear-axle ratio, vehicle weight and streamlining, etc., the vehicle designer has to arrange for the engine to expend fuel as far as possible in the region of best efficiency (figures 2.15 and 2.16).

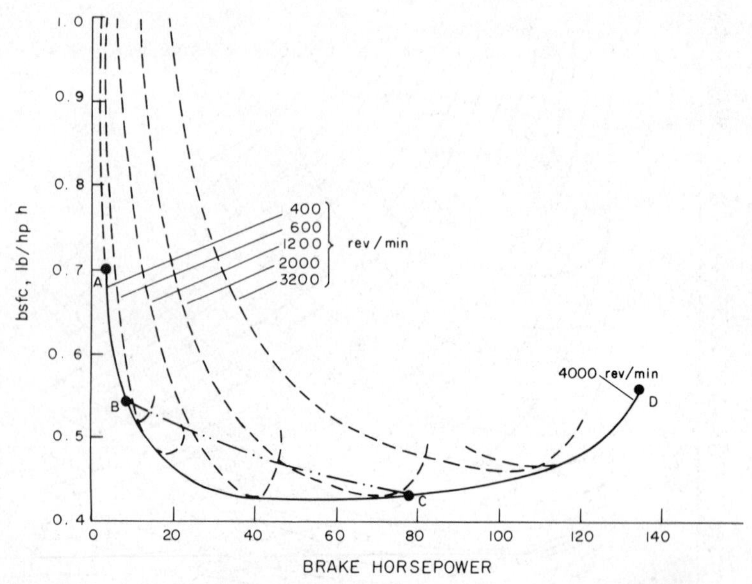

Figure 2.16 Typical variation of bsfc with power output at different engine speeds, showing strong increase at low loads due to pumping losses and small increases at high loads due to heat losses and mixture enrichment[17]: full curve ABCD shows the 'best economy' tune for this engine; chain curve BC shows the road-load locus of points (1 lb/hp h = 0.169 kg/MJ; 1 hp = 0.745 kW)

2.5.1 Transmission Efficiency and Design Ratios

It has long been known that significant benefits in fuel economy can be achieved by making modifications to the gearing. Caris and Richardson[17] in 1953 stated that 'the transmission should in effect act as a moderator to keep the engine always operating at its best efficiency to develop the required power'. Figure 2.17, taken from their paper, shows very clearly that, if an ideal and continuously variable gear ratio were available, the improvement in fuel economy over that obtained using a standard constant gear ratio would range from 70 to 80% over the speed range from 30 to 70 mile/h (from 48.3 to 112.6 km/h) for this engine. However, it has proved too expensive to build such an ideal transmission, and so compromises have been necessary with the result that a choice between maximizing power or economy has had to be made. This is illustrated very clearly in figure 2.18, where the trade-off between fuel consumption and acceleration performance is shown for a simulated vehicle of constant weight and engine size with various chosen axle ratios[18]. A minimum in economy is reached with an axle ratio of around 2.25 in this example. Any further decrease in axle ratio causes a loss in both fuel economy and performance, probably owing to poor combustion mixtures at the low engine speeds and large throttle openings involved, which necessitate carburetter enrichment and consequent loss of efficiency.

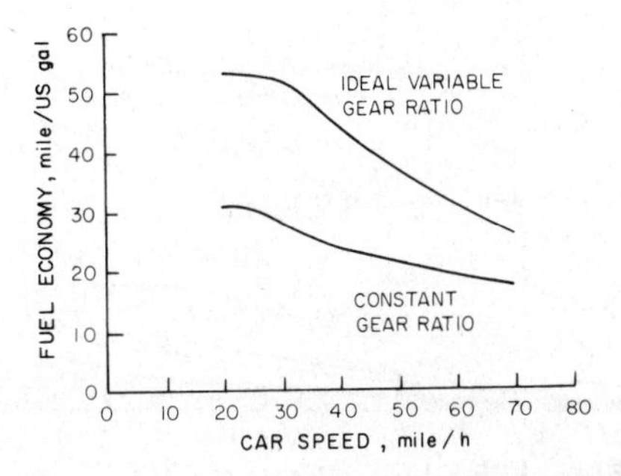

Figure 2.17 Variation of fuel economy with car speed, showing benefit to be gained from an ideally matched variable gear ratio compared with a typical constant value with a research engine in a stock car[17] (1 mile/US gal = 0.425 km/l; 1 mile/h = 1.609 km/h)

In practice there are, of course, also direct losses in efficiency involved in any gearbox, rear-axle or transmission system, but these are not very large for rear-axle gears or manual gearboxes. Automatic transmissions, which can potentially show fuel economy gains because of their torque multiplication and consequent permitted use of lower numerical rear-axle ratios, nevertheless also show power losses, due to friction bands, hydraulic clutches and pumps and high slippage[19] (figure 2.19). It appears that other benefits of automatic gearboxes (such as better low-speed torque and smoother operation, both of which may permit retuning of the engine for greater economy) are not sufficient to outweigh these high power losses, even with torque lock-up at high speeds.

Automatic transmission lubricants also have some effect on the efficiency of a vehicle, and this is discussed in chapter 11.

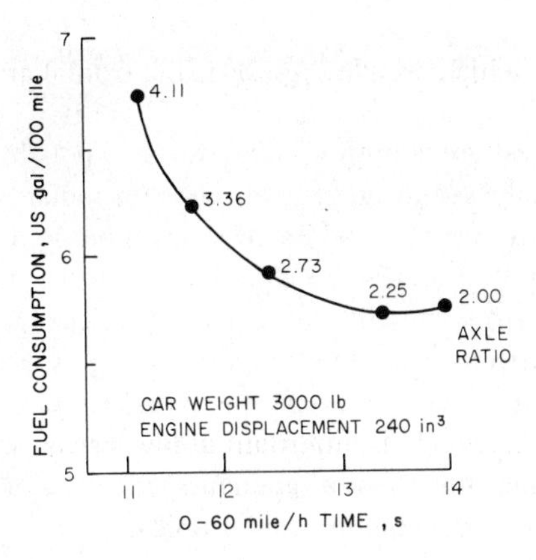

Figure 2.18 Trade-off between fuel consumption and acceleration performance with different axle ratios[18] (1 US gal/100 mile = 2.352 l/100 km; 60 mile/h = 96.5 km/h; 1 in^3 = 0.0164 l)

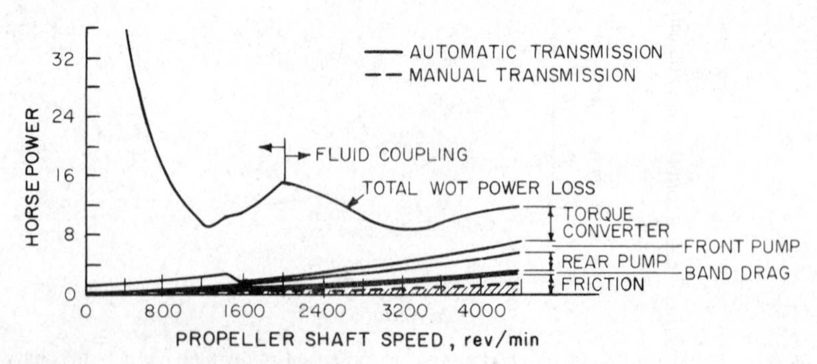

Figure 2.19 Variation of transmission power losses with speed in automatic and manual cars at WOT, showing how the torque convertor dominates the picture, particularly at low speeds owing to torque multiplication[19] (1 hp = 0.746 kW)

2.5.2 *Vehicle Road Load*

The tractive force required at the rear wheels to drive a vehicle at a steady speed is conveniently expressed by

$$\text{tractive force, lbf} = \frac{\text{bhp} \times 375}{v} = C_R W + 0.0026 C_D A v^2 + GW$$

where C_R and C_D are the coefficients of rolling resistance and drag, W is the gross vehicle weight, lbf, A is the frontal area, ft^2, v is the speed, mile/h, and G is the gradient.

In SI (metric) units

$$\text{tractive force } N = \frac{\text{brake power, kW} \times 3600}{\text{speed, km/h}} = 9.808 C_R M + 0.0982 C_D A' S^2 + 9.808 GM$$

where M is the gross vehicle weight, kgf, A' is the frontal area, m^2, and S is the speed, km/h.

The tyre rolling resistance term $C_R W$ is, as the expression indicates, linearly proportional to vehicle weight, in general is less for radial tyres ($C_R = 0.014$) than for cross-ply tyres (0.019) and decreases as the tyre pressure is increased.

The aerodynamic drag coefficient is commonly in the range 0.3–0.5, and, because this term is proportional to v^2, small design changes in vehicle body shape will very sensitively affect fuel economy at high speeds.

Vehicle weight enters the steady-speed expression twice: once in the rolling resistance term, which is important at low speeds, and again strongly in the term that describes the effect of gradients. Of course, another point where vehicle weight affects fuel economy very strongly is by virtue of the acceleration equation

$$\text{tractive force} = W \times \text{acceleration}$$

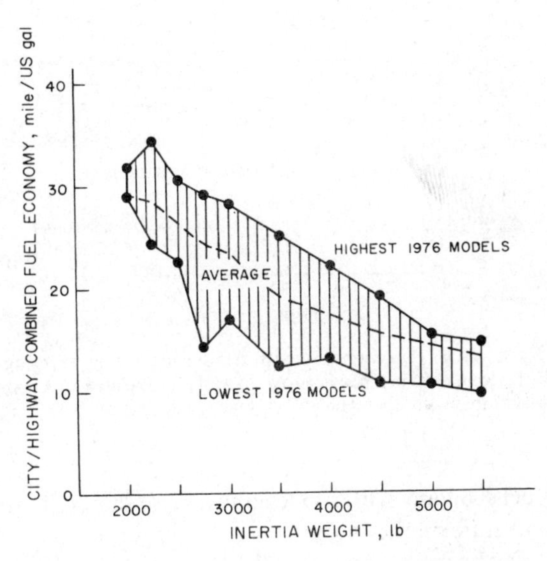

Figure 2.20 Variation of vehicle urban fuel economy with inertia weight for 1976 US cars[20] (1 mile/US gal = 0.425 km/l; 1 lb = 0.454 kg)

In mixed urban driving, vehicle weight accounts for a major part of the fuel usage, and probably this explains why this is the overwhelming factor that dominates recent US Environmental Protection Agency (EPA) fuel economy test data[20] (figure 2.20). There has been an inexorable trend to larger vehicles over recent years, and this has been helped by requirements for greater safety and control of exhaust emissions, in addition to features which gave the customer greater comfort or convenience.

The above considerations of vehicle road load show why it is practically so difficult to measure fuel economy on the road. Driving patterns, with different amounts and rates of acceleration, at different speeds and over different terrain, can vary so much from day to day or driver to driver that it is extremely difficult to get repeatable figures. On top of this, there is a strong effect of ambient conditions (temperature mainly, though wind, rain and sun all play their part) and an even stronger one of trip length as influenced by cold start (see below). These factors, and some of the techniques available for fuel economy measurement on the road and on a chassis dynamometer, are discussed more fully in chapter 9 and chapter 12.

2.5.3 Cold Start

It is becoming an increasingly well-recognized fact that cold-start fuel economy has an extremely damaging affect on a car's overall fuel economy. Data on US driving patterns indicate that as much as 50% of all gasoline is consumed in trips of 10 miles or less. Figure 2.21 shows[21] the dramatic

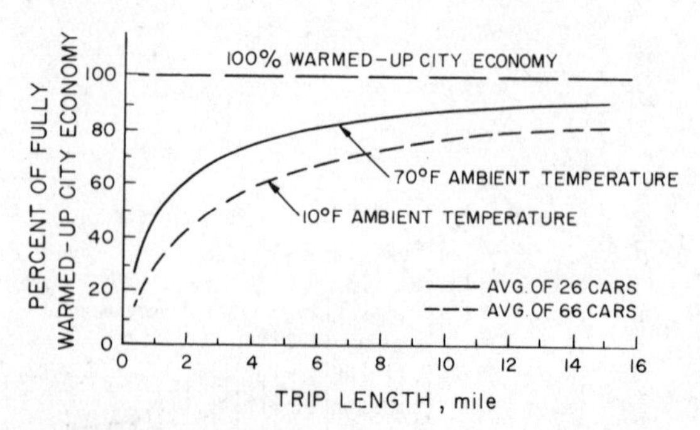

Figure 2.21 Variation of fuel economy with trip length during warm-up from cold start (showing both the long distance needed to achieve fully warmed-up economy and the marked effect of ambient temperature)[21] (1 mile = 1.609 km)

variation of fuel economy with trip length and also shows the strong effect of ambient temperature on the curve.

Attempts may be made to improve the above situation by designing a faster-warming engine or at least one that can accept a faster choke release pattern. Attempts to do the latter have been under way for some time with a view to controlling exhaust emission of CO in particular, but a balance has had to be struck between loss of driveability and emissions reduction. At this point an opportunity exists for fuel design to help, for it is possible to tailor volatility to help to achieve faster choke release rates without loss of driveability. There is also the possibility of a small direct beneficial effect by using fuels with improved volatility that give improved driveability (e.g. less hesitation, stumbling and stalling under cold-start conditions). These economy-enhancing gasolines are further discussed in chapter 4.

An opportunity also exists for the engine lubricant designer to improve this cold-start fuel economy, even though there may be less to choose between lubricants under fully warmed-up conditions. Tailoring of the viscosity–temperature relationship (under both static and dynamic conditions) can bring significant benefits, and these are further discussed in chapter 10.

2.5.4 Accessories

The power consumed in a modern vehicle by its accessories[22] has come to be considerable: the generator (0.5–2 hp), fan (0–7 hp), power steering (0.5–2.5 hp), power brakes, electric comfort gadgets (window heaters, power seats), as well as automatic transmission (10 hp) and air conditioning (11 hp) in hot climates all add up to a considerable total. In terms of their effect on urban fuel economy, their detrimental effects have been summarized[23] for US cars as follows: air conditioning, 7.7%, 9% and 13%; automatic transmission, 0–6%, 5–6% and 14–15%; alternators, 7.8%; fan, 1%; power steering, 1%.

Thus far, the customer has considered them to be worthwhile, but the increasing cost of energy may well cause a revision of the trend.

2.5.5 Emission Controls

The last decade has seen the growth of emission control technology, most notably in the US but also to an increasing extent in Europe. This has meant that vehicles have had to be redesigned somewhat and that the engines have been retuned. One notable feature that took place in the early 1970s was the re-introduction of low compression ratios (8:1 or less). Various estimates of the fuel economy penalties of all these changes have been made, and values of up to 20% were given. It is interesting to note that with the recent (1975) advent of exhaust oxidation catalysts, fuel economy benefits of around 20% have accrued[20], probably largely owing to the returning of the engine spark advance to its optimum position. This topic is more fully discussed in chapter 8, and the very interesting relationship between compression ratio and fuel economy when the refinery economics of high-octane gasoline manufacture are treated as a 'total system' are discussed in chapter 3.

2.6 What is the Potential for Fuel Economy Gain?

This very pertinent question is yet again being asked by vehicle designers as well as legislators, yet this time it is being asked against a background of increasingly tight emission control and safety regulations. A recent US official report to Congress by DOT–EPA[24] breaks the question down into different car sizes. They conclude that the largest single increase over 1974 fuel economy performance could come from engine modifications (15–25%), that about 9% could come from the use of four-speed automatic transmissions with lock-up and that 8–12% improvement could come from reduction of car weight, drag, rolling resistance and accessory power consumption. For cars above compact size, a further 10–15% could be gained from reducing the size of engines, and by 1980 a change in model size mix by the public could result in another 9% reduction.

To what extent these gains can be enjoyed with European-type cars is a debatable question. Certainly smaller gains would appear to be possible from engine size and model mix changes, since European cars are already smaller and more homogeneous. Wastefully geared automatics are also relatively scarce, so improvements in gearing should be less. However, engine modifications, combined with the trading of power for fuel economy by tuning modifications, should yield a significant benefit, though again this will most likely be smaller than the US predictions since compression ratios have not been decreased to the same extent nor have spark timings been so much retarded for emission control purposes. Nevertheless, engine efficiency gains should be realizable by attention to mixture strength, compression ratio, pumping losses and spark timing, the magnitude of the gains decreasing in the order given.

The author thus offers as a personal speculation of what is practically realizable over the next decade the following fuel consumption gains in European gasoline-powered cars.

(1) From engine design changes, 20%.
(2) From gasoline design changes, 5–10% according to driving conditions.
(3) From engine lubricant changes, about 3%.
(4) From transmission design changes, 5–10%.
(5) From transmission lubricant changes, about 3%.
(6) From design changes of a given vehicle (weight, drag, tyres, accessories), approximately 10%.
(7) From engine size and model mix changes, approximately 10%.
(8) From vehicle maintenance procedures, approximately 5%.

Since these different effects are largely independent of one another, the surprisingly high total emerges of about 50% potential improvement, which makes a very good target for the industry to aim at. It is not the first time that such an optimistic forward look has been taken. No less a person than Charles Kettering[25] when General Motors president in 1929 predicted '80 mile/gal by 1939'. Let us hope that, with the present depletion of oil reserves, the progress towards achieving the target will be faster in the coming decade than it was then.

References

1. P. F. Everall. The effect of road and traffic conditions on fuel consumption. *Min. Transport, Road Res. Lab. Rep.*, No. LR 226 (1968)
2. Shell. Unpublished work (1971)
3. F. D. Rossini et al. Selected values of physical and thermodynamic properties of hydrocarbons and related compounds. *Am. Pet. Inst. Res. Project*, No. 44, Carnegie Press, Pittsburgh (1953) and later reports from the Carnegie Institute of Technology
4. J. P. Pirault. Notes on bottoming cycles using exhaust heat of automotive engines. *Ricardo Consult. Eng. Shoreham-by-Sea, Sussex. Pap.*, No. DP 18268 (1974)
5. C. F. Taylor. *The Internal Combustion Engine in Theory and Practice*, 2nd edn, M.I.T. Press, Cambridge, Massachusetts (1966)
   C. F. Taylor and E. S. Taylor. *The Internal Combustion Engine*, International Texbook Co., Scranton, Pennsylvania (1966)
6. E. F. Obert. *Internal Combustion Engines*, 3rd edn, International Texbook Co., Scranton, Pennsylvania (1968)
7. L. C. Lichty. *Combustion Engine Processes*, McGraw-Hill, New York (1967)

8. H. Ricardo and J. G. G. Hempson. *The High-speed Internal Combustion Engine,* 5th edn, Blackie, London and Glasgow (1968)
9. M. H. Edson and C. F. Taylor. The limits of engine performance – comparison of actual and theoretical cycles. *Soc. Automot. Eng. Tech. Pap*., No. 7, *Digital Calculations of Engine Cycles* (1964) p. 65
10. J. J. Gumbleton, R. A. Bottom and H. W. Lang. Optimizing engine parameters with exhaust gas recirculation. *Soc. Automot. Eng. Pap.,* No. 740104 (1974)
11. D. F. Caris and E. E. Nelson. A new look at high-compression engines. *Soc. Automot. Eng. Pap.*, No. 61A (1958)
12. I. N. Bishop. Effect of design variables on friction and economy. *Soc. Automot. Eng. Pap.*, No. 812A (1964)
13. A. E. Cleveland and I. N. Bishop. Fuel economy. *Soc. Automot. Eng. Pap.*, No. 150A (1960)
    Also in *Soc. Automot. Eng. J.,* **68** (August 1960) 27
14. W. S. Jones. Some factors in gasoline economy. *Soc. Automot. Eng. Quart. Trans.,* **3** (1944) 516
15. *Du Pont Tech. Memo.*, No. 8026 (December 1974)
16. F. M. Strange. An analysis of the ideal Otto cycle. *Soc. Automot. Eng. Tech. Pap.*, No. 7, *Digital Calculations of Engine Cycles* (1964) 92
17. D. F. Caris and R. A. Richardson. Engine–transmission relationship for higher efficiency. *Soc. Automot. Eng. Trans.,* **61** (1953) 81
18. C. Marks and G. W. Niepoth. Car design for economy and emissions. *Soc. Automot. Eng. Pap.*, No. 750954 (1975)
19. W. E. Zierer and H. L. Welch. Where does all the power go? – III. Effective power transmission. *Soc. Automot. Eng. J.,* **65** (April 1957) 54
20. T. C. Austin, R. B. Michael and G. R. Service. Passenger-car fuel economy trends through 1976. *Soc. Automot. Eng. Pap.*, No. 750957 (1975)
21. C. E. Scheffler and G. W. Niepoth. Customer fuel economy estimated from engineering tests. *Soc. Automot. Eng. Pap.*, No. 650861 (1965)
    Also in *Soc. Automot. Eng. J.,* **74** (1965) 46
22. E. C. Campbell. Where does all the power go? – II. The accessories – The first bite. *Soc. Automot. Eng. J.,* **65** (April 1957) 54
23. Various authors quoted by T. C. Austin and K. H. Hellman. Passenger-car fuel economy – Trends and influencing factors. *Soc. Automot. Eng. Pap.*, No. 730790 (1973)
24. Potential for motor vehicle fuel economy improvement. *DOT–EPA Rep.,* submitted to the US Congress (October 1974)
25. Interview by H. M. Robinson. *Pop. Sci. Mon.* (September 1929)
26. Shell. Unpublished work (1966)

3 Motor Gasoline and the Effect of Compression Ratio on Octane Requirement and Fuel Economy

A. G. BELL

3.1 Introduction

The purpose of this chapter is to provide the general reader with an overall review of the principal characteristics of a motor gasoline, to study in some detail the effect of compression ratio on octane requirement and fuel economy and then to look at the overall economy when the car and the refinery are considered as a single economic unit.

Those readers who are already familiar with the basic characteristics of gasoline and how they affect performance may wish to skip lightly over section 3.2.

3.2 Motor Gasoline and its Effects on Performance

The principal requirements of a motor gasoline are the following.

(a) It should be sufficiently volatile to give inflammable mixtures under all operating conditions.

(b) It should burn smoothly without knock.

(c) It should give good fuel economy.

(d) It should keep inlet system and combustion chamber deposits to a minimum.

(e) It should burn cleanly and should help to keep pollution of the atmosphere at a low level.

3.2.1 Volatility

Motor gasolines consist essentially of mixtures of hydrocarbons boiling in the range from about 30 to 200 °C, and the volatility of a gasoline is usually

Table 3.1 Distillation characteristics of a volatile and a less-volatile gasoline

Temperature, °C	Percentage (volume) distilled	
	Volatile	Less volatile
48	18	1
70	42	10
100	70	38
120	82	55
160	98	80
170	100	87
210		100

expressed in terms of the volume percentage that distils over at, or below, certain fixed temperatures. Distillation figures for two extreme gasolines are given in table 3.1.

The differences between these gasolines can perhaps be shown more clearly by plotting what are known as the distillation curves of the gasolines, as in figure 3.1; the distillation curves for the majority of commercial gasolines lie between the two shown.

Most gasolines contain small quantities of gaseous hydrocarbons such as butane, and the quantity of these gases is usually controlled by measuring

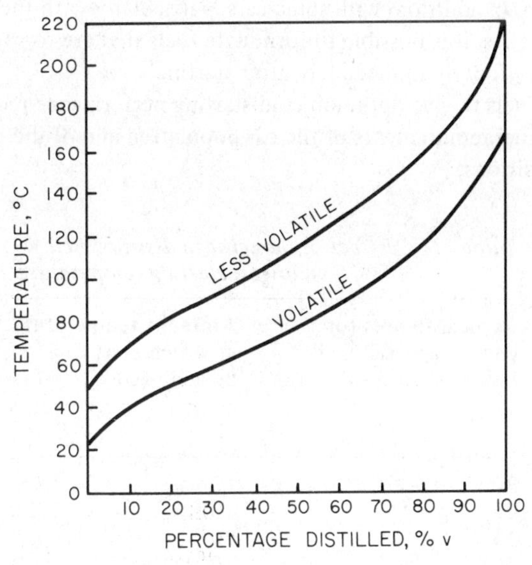

Figure 3.1 Distillation curves for gasolines listed in table 3.1

what is known as the Reid vapour pressure of the gasoline. The greater the butane content, the higher is the Reid vapour pressure.

3.2.2 Fuel Volatility and its Effect on Cold Starting

When an engine is cold, only part of the gasoline in the inlet manifold vaporizes, and to ensure that an inflammable mixture reaches the cylinders extra fuel is supplied by means of the choke or the starter carburetter. The amount of extra fuel required depends on the ambient temperature and the volatility of the gasoline – the more volatile the gasoline, the more of it vaporizes under any given condition and the smaller is the amount of extra fuel required.

In practice it is the more volatile parts of the gasoline, as characterized by the percentage distilling below 70 °C, that influence the ease with which engines can be started; the greater the percentage that distils below 70 °C, the better is the cold-starting performance. This effect is illustrated in table 3.2 which shows how the temperature at which a satisfactory start, i.e. start in less than 5 s using full choke, varies with the percentage distilled below 70 °C.

In practice, motorists generally prefer their cars to start more easily than this, if possible without the use of the choke, and for this, and other reasons that will be discussed later, most marketed gasolines are more volatile than would appear to be absolutely necessary on the basis of the figures in table 3.2. The upper limit of volatility, i.e. the maximum percentage distilling below 70 °C (and the maximum allowable Reid vapour pressure), is governed by the tendency for more volatile gasolines to give rise to vapour lock (see section 3.2.3). In addition, with some cars, particularly with those fitted with automatic chokes, it is possible to formulate fuels that are too volatile for the engine to run properly immediately after starting.

Blending fuels to give optimum cold-starting performance requires a knowledge of the fuel requirements of the car population and of the expected ambient conditions.

Table 3.2 Effect of percentage distilled below 70 °C on lowest starting temperature

Typical figures for percentage distilled below 70 °C, %v	Ambient temperature at which start can be achieved	
	°F	°C
30	0	−17.8
25	10	−12.2
15	20	−6.7
10	30	−1.1

3.2.3 Hot Fuel Handling

Light gases, such as butane, are dissolved in motor gasoline to give good starting. If, for any reason, the fuel becomes hot, then these gases and other light hydrocarbons tend to boil off and to form slugs of vapour in the fuel system. This may prevent the fuel pump or the carburetter from functioning properly and may cause engine roughness and loss of power, or may prevent the engine from being restarted after a short shut-down.

By providing adequate under-bonnet cooling, by keeping fuel supply lines away from the exhaust pipe or by putting the fuel pump in or near the fuel tank, manufacturers have been able to reduce this tendency. These mechanical solutions to the problem often run counter to other car-manufacturing pressures, particularly cost, styling and the allocation of more and more under-bonnet space to accessories. As a result, vapour locking of the fuel pump and allied carburation difficulties continue to be problems which would affect many cars were it not for the imposition of limits on the Reid vapour pressure and the percentage distilled below 70 °C.

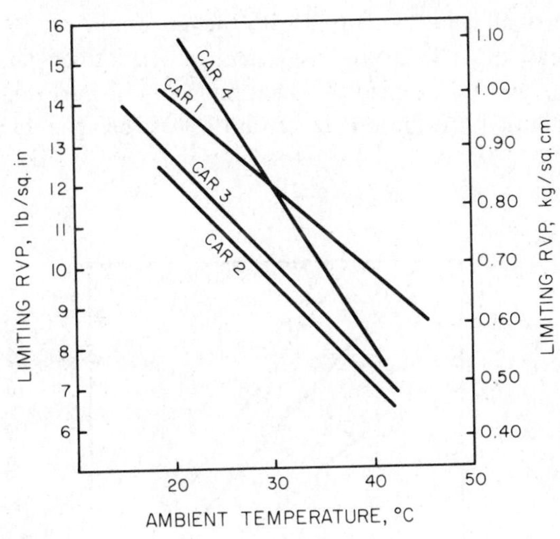

Figure 3.2 Effect of ambient temperature on maximum Reid vapour pressure (RVP) for four types of car

The limiting Reid vapour pressure depends on the type of engine and the ambient temperatures expected during the period over which the gasoline will be sold. Figure 3.2 shows the limiting Reid vapour pressure for four different makes of cars at various ambient temperatures.

The problems of hot fuel handling are aggravated by altitude, and fuels to be sold at the higher altitudes are blended to compensate for this effect.

3.2.4 *Warm-up*

During the first few minutes after a cold start an engine is sluggish because the inlet manifold has not yet reached its working temperature and because the engine, gearbox and back-axle oils are cold. Supplying extra fuel by means of a starter carburetter or a choke can improve the performance somewhat, but this is wasteful and can lead to excessive dilution of the engine oil by unburnt fuel and to increased exhaust emissions.

The use of exhaust- and water-heated inlet manifolds reduces the length of the warm-up period but, unless carefully controlled, can also reduce the maximum power available when the engine is fully warmed up.

Increasing the volatility of the gasoline also reduces the length of the warm-up period, and it has been found that the controlling factor in this case is the mid-range volatility as characterized by the percentage that distils below 100 °C. Figure 3.3 illustrates this effect and shows how in a particular car, at an ambient temperature of 0 °C, the use of a fuel with 42%v evaporated at 100 °C gives a warm-up distance of 4.3 km, whereas with a fuel with 60%v distilled at 100 °C the warm-up distance is only 2.5 km. Some cars are more and others less sensitive to changes in the percentage distilled at 100 °C – the least sensitive of all being cars fitted with fuel injection.

As in the case of cold starting, the extent to which the volatility can be increased, and hence the extent to which the warm-up performance can be improved, is limited, the limiting factor in this case being carburetter icing.

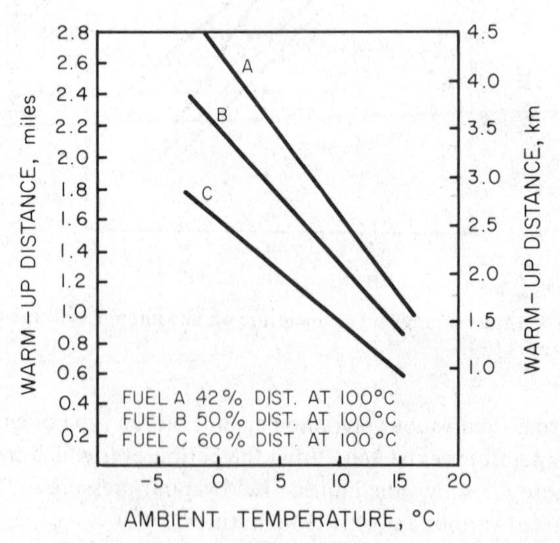

Figure 3.3 Effect of changes in percentage distilled at 100 °C on the warm-up distance (warm-up distance is the distance to reach 70% fully warmed-up performance)

3.2.5 Carburetter Icing

When fuel vaporizes in the carburetter, heat is drawn from the carburetter, and its temperature falls. On a cold damp day this reduction in temperature causes moisture from the air to condense on the carburetter in the form of ice. This icing-up of the carburetter may cause the engine to stall or idle roughly when the car is held up in traffic, or it may even bring the car to a complete standstill and may prevent it from being restarted until the ice has melted.

The design of the carburetter has a marked influence on the severity of carburetter icing; the better the atomization of the fuel the more the carburetter is likely to suffer from icing; this is unfortunate as good atomization is desirable from other points of view. It is, therefore, becoming increasingly common practice for motor manufacturers to fit devices which enable the carburetter to pick up warm air from near the exhaust pipe during cold weather; this reduces the problem but does not eliminate it entirely.

The more volatile a gasoline, the more of it evaporates in the carburetter and the greater the temperature drop. Thus, making a gasoline less volatile helps to prevent carburetter icing, but, as we have already seen, it makes the warm-up performance worse. Fortunately, a number of chemical compounds, such as isopropyl alcohol and some glycols and amines, reduce carburetter icing, and these can be included in winter-grade gasolines for this purpose. The use of these anti-icing additives, as they are called, permits volatile gasolines to be used during the winter months so that warm-up distance is kept to

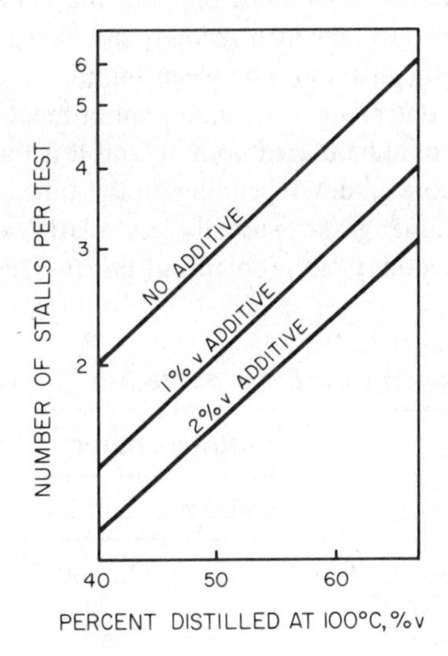

Figure 3.4 Effect of percentage distilled at 100°C and an anti-icing additive on carburetter icing

a minimum and yet the formation of ice is prevented. These additives are, however, expensive and to achieve optimum performance in the most economical manner requires skilled blending and careful consideration of expected weather conditions.

This, together with the fact that there is available to the refiner a wide range of additives of varying effectiveness, can lead to considerable variations in performance between competitive gasolines in this respect.

Figure 3.4 shows how in a critical car the number of engine stalls occurring in the first few kilometres of stop–start driving is influenced by the percentage distilled at 100 °C and by the presence of an anti-icing additive.

3.2.6 Fuel Economy

Detailed descriptions of how various engine and fuel factors affect the overall economy of an engine are described in section 3.3 and in chapters 4, 5 and 6.

3.2.7 Overall Volatility Considerations

The aim of the gasoline blender is to blend the gasoline to give good cold starting, warm-up performance and fuel economy, without giving rise to carburetter icing or problems of hot fuel handling.

As will be clear from the preceding sections, this is not an easy task, and the ultimate composition of the gasoline will depend on the ambient temperature expected in the area in which the gasoline will be sold and the type of car in the market. This in turn requires a continuous monitoring of the fuel requirements of new cars as they come onto the market and constant review of the fuel specifications to take into account any changes.

In many countries the difference between summer and winter temperatures are so great that one grade of gasoline cannot meet the conflicting requirements which would be placed upon it, and it is therefore necessary to market two or even more grades depending on the time of year. In these circumstances, the summer grade is usually less volatile, since cold starting is not a problem and freedom from problems of hot fuel handling plus good fuel

Table 3.3 Typical volatility characteristics for various areas

	Northwest Europe		Central Africa
	Winter	Summer	
Reid vapour pressure, lb/in^2	13.5	10.0	7.5
%v distilled at 70 °C	35	25	10
%v distilled at 10 °C	50	45	38
%v distilled at 160 °C	95	80	80

economy are paramount, whereas the winter grade is more volatile so as to provide good cold starting and warm-up and may contain anti-icing additives to prevent carburetter icing.

Table 3.3 illustrates how Reid vapour pressure and distillation characteristics vary from country to country, and season to season, depending on ambient temperature.

3.2.8 Anti-knock Performance

The higher the compression ratio of an engine, the better is its overall efficiency and the higher the maximum obtainable power output.

The extent to which the compression ratio can be raised is limited by the onset of knock and this, in turn, depends on the anti-knock quality of the fuel.

The anti-knock quality of a fuel is expressed in terms of its octane number and is measured by comparing the performance of the fuel against the performance of mixtures of iso-octane (100-octane number) and n-heptane (0-octane number) in a standard laboratory test (a gasoline of 98-octane number is one which gives the same anti-knock performance as a mixture of 98%v of iso-octane and 2%v of n-heptane). The actual octane level required for any particular engine depends both on its compression ratio and on certain mechanical design features. An indication of how the octane requirement of one particular engine varies with compression ratio is shown in figure 3.5.

The octane numbers given in this figure are research octane numbers (RONs); these are measured under relatively mild test conditions. Under more

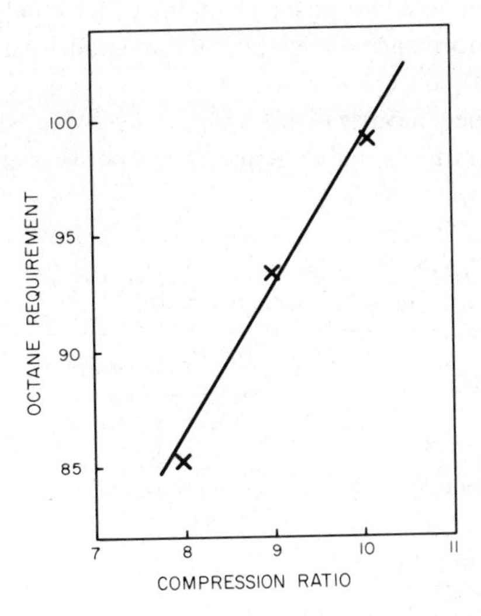

Figure 3.5 Octane requirement versus compression ratio

Table 3.4 Effect of motor octane number on road performance

		Gasoline A	Gasoline B
RON		98.8	98.6
MON		84.8	91.7
Road octane number[a] (at constant speed)	car 1	96.0	93.8
	car 2	96.9	96.9
	car 3	91.1	100.5

[a]Road octane number is the anti-knock performance measured in the car on the road relative to the same mixtures of iso-octane and n-heptane.

severe test conditions some fuels do not give as good an anti-knock performance as their RONs would indicate, and it is therefore necessary to measure the performance of gasolines both by the research method and also by a more severe method, called the motor method, to give motor octane numbers (MONs).

In practice, operating conditions in a car on the road may be more or less severe than those in the RON test so that gasolines with low motor method ratings may give either better or worse performance than would be expected from their RON. Some values illustrating this are given in table 3.4, which shows that gasoline A is better in car 1 than gasoline B but worse in car 3. In car 2 both gasolines gave the same performance.

It has also been found that the RON of the more volatile portions of the gasoline (e.g. that fraction which distils below 100°C) is of importance in controlling anti-knock performance under accelerating conditions.

Table 3.5 shows how two gasolines of equal RON and MON can give very different road performance depending on the octane quality of the fraction distilling up to 100 °C.

Petroleum refiners have available a wide range of possible processes by which they can produce gasoline blending components and have a number of

Table 3.5 Effect of the octane quality of the light fractions of a gasoline on road performance

		Gasoline C	Gasoline D
RON		95.6	95.5
MON		85.5	85.4
RON fractions distilling below 100 °C		85.7	94.6
Road octane number (accelerating)	car 4	88.9	91.9
	car 5	91.2	93.5
	car 6	88.5	94.7

different anti-knock additives to choose from. By combining these components in different ways and by varying the percentages of different anti-knock additives used, gasolines with various RONs, MONs and RONs of the fraction distilling below 100 °C may be obtained. There are, however, many restrictions on the use of different blending components, and no two refiners will arrive at exactly the same combination of components and additives, and a discriminating motorist driving a critical car can, therefore, often detect the difference between two gasolines which have the same RON number, i.e. the motorist is able to distinguish between gasolines which differ in MON or in RON of the more volatile portion.

At the present time the required levels of octane quality are reached by a combination of refinery processing and the use of lead anti-knock additives. When the use of lead alkyls is prohibited in certain grades, as it has been in the US and in Japan, it is not generally possible to provide a premium-quality gasoline of the same octane number. Consequently the unleaded grades are usually of regular or intermediate octane quality. Even so, alternative high-octane components may have to be used instead of the lead, and these components may require new refinery processing plant, which involves time and money for construction.

Where the use of lead alkyls is restricted though not eliminated but where the octane number has to be maintained, as is happening in a number of other countries, then alternative high-octane components are likely to be essential.

In either case, whether it is no lead or low lead, the performance of the fuels may therefore be different as a result of the changed composition, and this in turn may necessitate changes in compression ratio and ignition timing to ensure that cars do not knock. The net effect of these changes on fuel manufacturing cost and overall vehicle economy is discussed in detail in section 3.4.

3.2.9 Cleanliness and Reliability

During the life of a car many thousands of litres of gasoline pass through the fuel system and carburetter to the combustion chamber, and, as a result, even small quantities of corrosive substances, solid impurities or gummy materials in the fuel can greatly reduce the overall life and general reliability of the engine.

Impurities and water are removed from the gasoline by settling and filtration, both before the fuel leaves the refinery and after its delivery to the depot. 'Good housekeeping', as it is called, is essential to the provision of clean uncontaminated fuel.

Freedom from gummy materials is ensured by the use of refinery processes which eliminate unstable materials, and the formation of gum during extended periods of storage is prevented by the use of anti-oxidants and metal de-activators.

In the past, adequate carburetter and inlet system cleanliness was achieved solely by preventing the formation of gummy materials, but the fitting of devices which return crankcase blowby gases to the inlet systems causes additional deposits in the inlet system and can lead to the production of richer fuel--air mixtures, giving increased exhaust emissions and poor fuel economy. Deposits on inlet valves can lead to power loss and valve sticking. It is therefore becoming increasingly common for fuel suppliers to add detergents to their gasolines, to keep carburetters and inlet systems free from such deposits.

The best of these additives are not only effective in preventing deposit formation but also give significant reductions in emissions and improvements in fuel economy. These effects are discussed in detail in chapter 5.

3.2.10 Exhaust Emissions

The exhaust gases from a gasoline engine contain not only traces of unburnt and partially burnt hydrocarbons from the fuel but also carbon monoxide, oxides of nitrogen and small quantities of lead from the combustion of lead anti-knock additives.

In areas of high traffic density and in areas where there is air pollution from other sources, the presence of these materials in exhaust gases can be objectionable.

In the US and Japan there are places where, for climatic and geographic reasons, air pollution presents a particular problem, and these countries have therefore introduced legislation calling for drastic reduction in the emission of carbon monoxide (CO), oxides of nitrogen (NO_x) and hydrocarbons (HC) by motor vehicles.

The use of moderately weak mixtures is only partially effective as a method of controlling emissions, and special techniques of mixture preparation and exhaust gas treatment are needed to meet the requirements. Exhaust CO and HC can be reduced by after-burners (thermal reactors) or catalytic reactors, and NO_x can be controlled by recirculating part of the exhaust gas to the inlet of the engine or by catalytic treatment.

In general gasoline composition has relatively little effect on exhaust emissions, but the presence of lead anti-knock additives would seriously reduce the life of many potential exhaust gas catalysts, and in both the US and Japan unleaded gasoline is provided for cars fitted with catalytic reactors.

In Europe where for climatic and geographic reasons air pollution is less of a problem and legislation is correspondingly less restrictive, there is no need to use catalytic reactors, and there is therefore no immediate need for the introduction of non-leaded gasolines.

It is, however, certain that the control of exhaust emissions presents a major technological problem, and the design changes made to reduce emissions can have a profound effect on fuel economy: this is discussed in detail in chapter 8.

3.3 Octane Number, Compression Ratio and Economy

As has already been seen in chapter 2 the thermal efficiency of a gasoline engine increases with increasing compression ratio. In practice this increase in efficiency is reflected in increased power output and improved brake specific fuel consumption (bsfc) (figures 3.6 and 3.7). The extent to which the improvement in bsfc will appear as improved economy on the road will depend on whether advantage of the increased power is taken to alter the axle ratio to give an additional improvement in economy or whether the axle ratio is held constant and the additional power is used to give improved acceleration (figure 3.8).

One of the most comprehensive studies of the effect of compression ratio on fuel economy was that carried out by Caris and Nelson[2] of General Motors in the late 1950s, and some of their principal findings are summarized in figures 3.9, 3.10 and 3.11. Figure 3.9 illustrates the effect of changes in both compression ratio and mixture strength and shows the considerable gains in brake thermal efficiency (bthe) that can be obtained from the use of

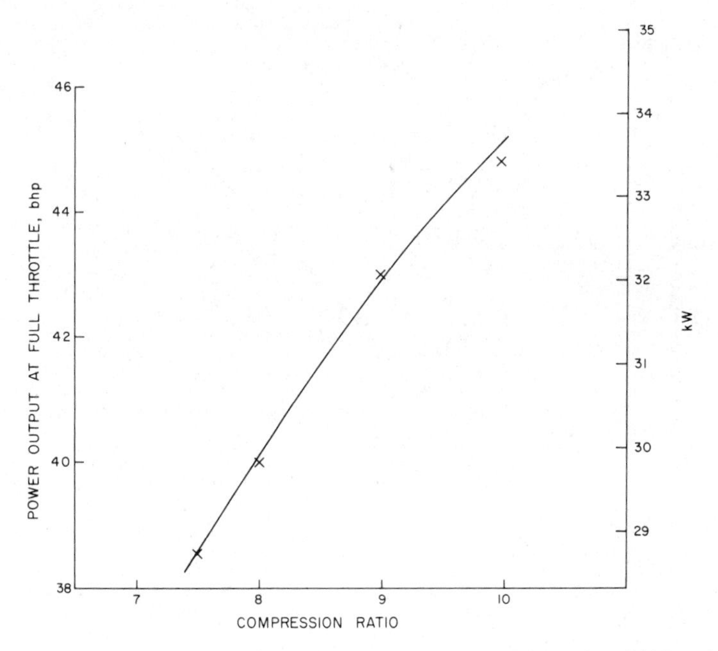

Figure 3.6 Effect of compression ratio on power output (2.2 ℓ engine, 2500 rev/min, constant speed)

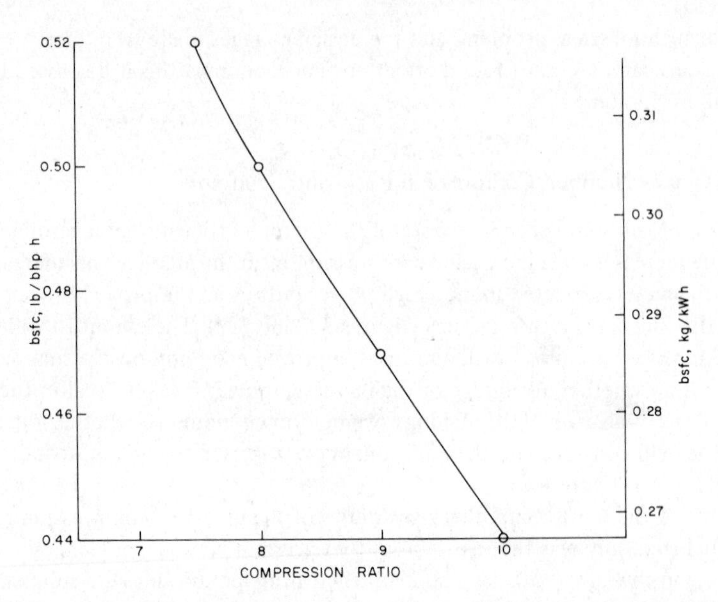

Figure 3.7 Effect of compression ratio on economy (2.2 l engine, 2500 rev/min, constant speed) (1 kg/kW h = 0.278 kg/MJ)

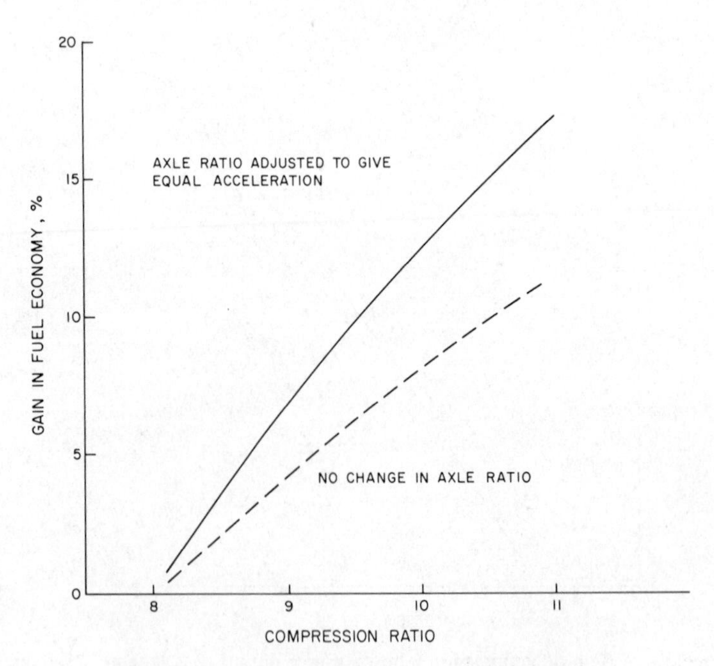

Figure 3.8 Effect of change in axle ratio on economy[1]

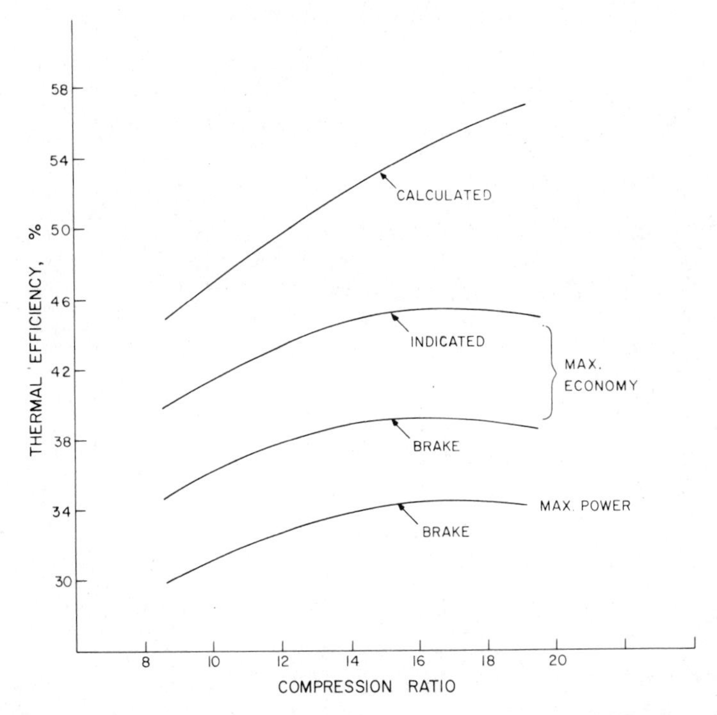

Figure 3.9 Effect of compression ratio on thermal efficiency at WOT and constant speed[2]

a combination of high compression ratio and maximum economy mixture strength. This figure also shows that increasing the compression ratio above about 16:1 or 17:1 led to a decrease rather than to an increase in efficiency. This was rather unexpected, and after a thorough investigation Caris and Nelson concluded that the reason for this was a decrease in the rate of combustion resulting from a high surface/volume ratio and generally poor combustion chamber configuration at the higher ratios. It is possible therefore that attention to combustion chamber design might prevent or delay the fall-off in efficiency at higher ratios.

Figures 3.10 and 3.11 illustrate the relatively poor thermal efficiency of the engine when running at part throttle. At first sight these would seem to argue an overwhelming case for the use of small engines operating at wide-open throttle (WOT) as opposed to the larger engine running at light load. It must be remembered, however, that, if the engine is too small, then it will have to run at maximum power mixture strength in order to produce the required power output and that, if in order to produce the necessary power it has to operate at high speed, then some of the economy benefit will be offset by other factors such as increases in engine friction.

On the basis of this information it seems not unreasonable to speculate

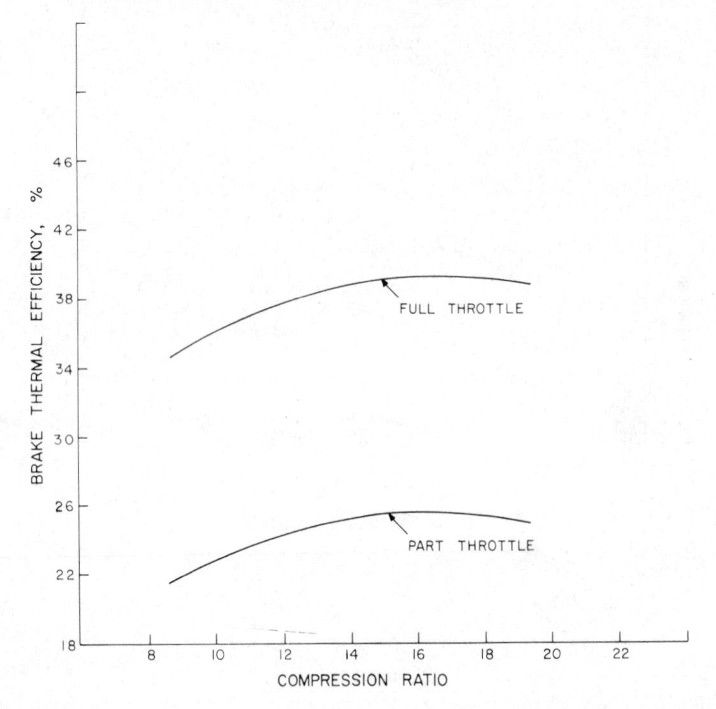

Figure 3.10 Bthe at full and part throttle (maximum-economy settings in both cases)[2]

that, if, in future, compression ratios are increased in the search for improved economy, these increases will be accompanied by changes in combustion chamber shape designed to improve the rate of combustion, by changes in mixture strength aimed at optimizing efficiency rather than power and by changes in axle ratio designed to make sure that the engine operates in the most efficient region under normal everyday driving conditions.

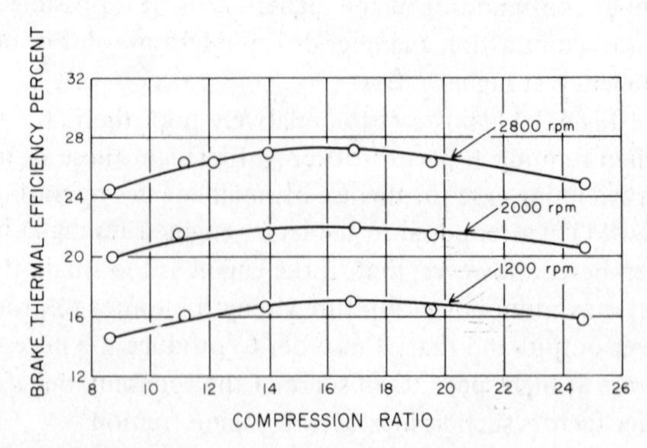

Figure 3.11 Effect of compression ratio on bthe under road-load conditions, at 30, 50 and 70 mile/h (48.3, 80.5 and 112.7 km/h) (same axle ratio in all cases)[2]

Some attempts have already been made in this direction[3], but at the time of writing it is not entirely clear whether such engines will gain general acceptance or whether they will be able to meet increasingly severe legislative requirements on exhaust emissions.

In the meantime, recent figures from the UK motor industry[4] (table 3.6) suggest that, for existing engines with compression ratios in the region of from 7.5 to 9.5:1, a unit increase in compression ratio can improve economy by from 4 to 12% with an overall average of 7%. This figure seems high in relation to the possible theoretical improvement and also in relation to earlier published work[1,5,6] and to the results of some limited testing at the Thornton Research Centre. All of these suggest an improvement of about 4% in economy per unit increase in compression ratio in the region of from 7.5 to 9.5:1. It may be that, in the case of the tests carried out by the UK manufacturers, changes in compression ratio were accompanied by other changes, e.g. in carburetter and ignition settings and that these contributed to the improvement.

It must be emphasized, however, that much of the published information on the effect of compression ratio changes on economy is based on the results of bench engine tests and that there is as yet relatively little good information on the effect on the road under normal everyday driving conditions. Nevertheless it is clear that, up to compression ratios of about 12:1, significant improvements in economy can be achieved and that smaller additional improvements may be obtained by going to even higher compression ratios.

In practice, however, the majority of engines are limited to much lower compression ratios by the onset of knock or other combustion abnormalities such as rumble and run-on. Typically, engines with compression ratios of 7.5:1 need gasolines of about 90 RON to prevent knock, and those with compression ratios of from 9.5 to 10.0:1 need fuels of from 99 to 100 RON. There are, however, very large differences in requirement between nominally identical engines and between engines of different design with the same compression ratio. For example, it can be seen from figure 3.12 that engines with compression ratios of 9.0:1 may have mean octane requirements ranging from 95 to 101 RON.

There are many reasons for such differences in requirement including differences in cylinder head design and differences in ignition systems and carburation. For example, if an engine knocks at low speed under accelerating conditions, the design of the inlet system is critical, for, if only part of the fuel is vaporized, there will be large differences in air/fuel ratio between the cylinders, and some may receive only the volatile fractions which may have a lower octane number than the rest of the fuel. Conversely, if an engine knocks at high speed, it is generally more critical of MON than RON, and the requirement is highly dependent on the air inlet and mixture temperature.

Nevertheless taken overall it seems that, for existing engines, unit increase in compression ratio generally raises the octane requirement by about four octane numbers. If one combines this figure with that given above of a 4%

Table 3.6 Test bed results[a] *(steady speed)*

Engine		Compression ratio range	Compression ratio span	Octane requirement range	Octane span	Octane number/compression ratio	Increase in fuel consumption, %v	
Code	Capacity, cm^3						Per unit drop in compression ratios	Per unit drop in octane ratios
A	1800	6.9–9.5	2.6	87–97[a]	10	3.8	–	–
B	1275	8.0–9.8	1.8	90–97	7	4.0	–	–
C	1275	7.9–9.4[a]	1.5	88–96[a]	8	5.3	6.0	1.1
D	1500	7.5–9.0	1.5	90–97	7	4.7	5.7	1.2
E	2000	8.0–9.3	1.3	90–97	7	5.6	7.3	1.3
F	2000	8.5–9.5	1.0	90–97	7	7.0	8.7	1.2
G	1600	7.5–8.7[a]	1.2	90–97	7	5.8	8.3	1.4
H	1300	7.8–8.7[a]	0.9	90–97	7	7.8	9.8	1.3
I	1498	7.5–9.6[a]	2.1	86–96[a]	10	4.8	3.4	0.7
J	1592	8.0–10.3[a]	2.3	90–98[a]	8	3.5	4.1	1.2
K	4235	7.6–10.0[a]	2.4	90–98[a]	8	3.3	5.1	1.5
L	1100	7.9–9.0[a]	1.1	90–97[a]	7	6.7	5.3	0.8
R	3442	7.7–8.4[a]	0.7	94.3–97[a]	2.7	3.9	12.0	3.1
S	4235	8.3–10.0[a]	1.6	97–97[a]	–	–	10.5	–

[a]Measured values – all others are nominal.

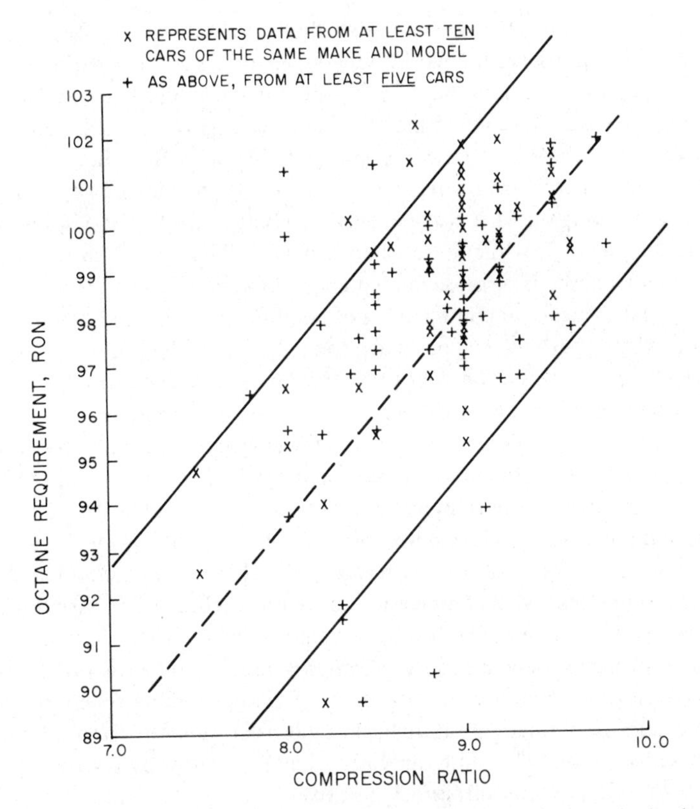

Figure 3.12 Effect of compression ratio on octane requirement (data obtained on commercial fuels of sensitivity 10 and $\Delta R_{100°C}$ 10; points indicate level at which 90% of cars of that model are satisfied)

improvement in economy per unit increase in compression ratio, one can conclude that, if in current engines the compression ratio is tailored to make best use of the available fuel, a unit increase in octane number will result in a 1% improvement in economy.

3.4 The Car and the Refinery as a Single Economic Unit

The earliest motor gasolines, which were produced as a by-product of kerosene manufacture, had extremely low octane numbers, and attempts by the car manufacturers to increase power output by increasing compression ratio were limited by the onset of knock. With the growth of the car industry the oil companies invested heavily in plant for the production and upgrading of motor gasoline, and for more than half a century the development of both industries has gone hand in hand.

The sudden cessation of supplies of crude oil from the Middle East in the autumn of 1973 and the cost escalation that followed has focussed attention on the need to conserve the world's energy resources and has led to a number of studies of the optimum use of crude oil. Some of these have included far-reaching proposals that would include major redesign of our cities and the replacement of the passenger car by public transport. Others have considered the possible replacement of the gasoline engine by the diesel or by some form of hybrid running on wide cut fuel encompassing much of the gasoline and diesel fuel range[7,8]. While these schemes may well have merit in the longer term, what is perhaps of more immediate short-term concern is the question of what is the optimum combination of compression ratio and octane number for existing gasoline engines when looked at from the point of view of overall energy consumption – both on the road and in the refinery.

In principle there is no difficulty in answering the question, as it is well known that increasing the compression ratio improves economy over the road but requires the use of higher-octane gasolines which in turn require more energy to produce in the refinery. In practice it is difficult to reach a generally applicable universal conclusion because, not only are there wide variations from one car model to another in terms of the changes in economy and octane requirement with compression ratio, but there are also variations from refinery to refinery in the additional amount of energy needed to produce gasolines of higher-octane quality. Furthermore, both the effect of compression ratio on economy and the effect of octane number on energy consumption in the refinery are influenced by legislation – the former by limits on the emission of NO_x and hydrocarbons and the latter by restrictions on the use of lead alkyls as anti-knock additives.

The question of the effect of NO_x and hydrocarbons on the relationship between compression ratio and fuel economy is a particularly difficult one in that General Motors have shown that the relationship is affected not only by the limits imposed but also by the methods used for emission control[9,10].

In Europe, where at the present time emission legislation is not so severe, van Gulick[11], working on the assumption that an increase in octane number of one would permit an increase in compression ratio that would give a 1% improvement in gravimetric fuel consumption on the road, has calculated the joint effect of increasing compression ratio and octane number on refinery crude oil consumption, operating cost and capital investment. The results of these calculations are summarized in figures 3.13, 3.14 and 3.15 which show the following.

(a) A combination of high compression ratio and high octane number offers a small overall saving in energy in the order of 2 tons of crude for every 100 tons of gasoline for an increase in octane number from 92 to 96. (As in Europe only about 20% of the barrel is sold as gasoline the overall saving in terms of total crude consumption would be less than ½%.)

(b) Minimum manufacturing cost is achieved at about 96–97 octane

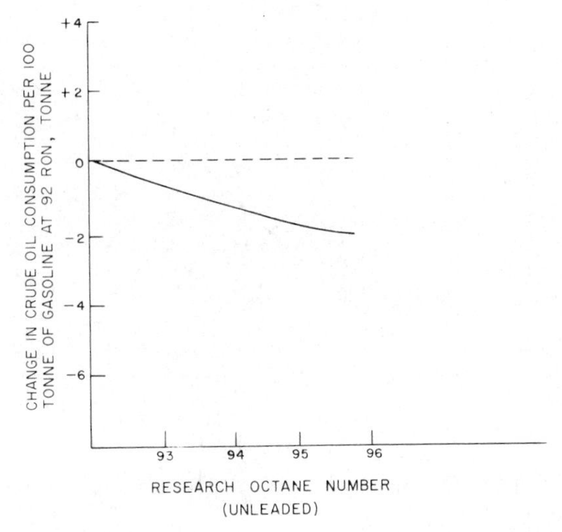

Figure 3.13 Effect on crude oil consumption of going to higher octane number/higher compression ratio (assuming 1% improvement in economy per octane number)[11]

number for leaded gasolines and at 92–93 octane number for non-leaded gasolines (figure 3.14).

(c) The capital cost of providing the plant necessary to make high-octane-number gasolines available is high even by oil industry standards (figure 3.15).

It can also be shown that the high-compression-ratio high-octane-number route will almost always be favoured by the individual motorist because his

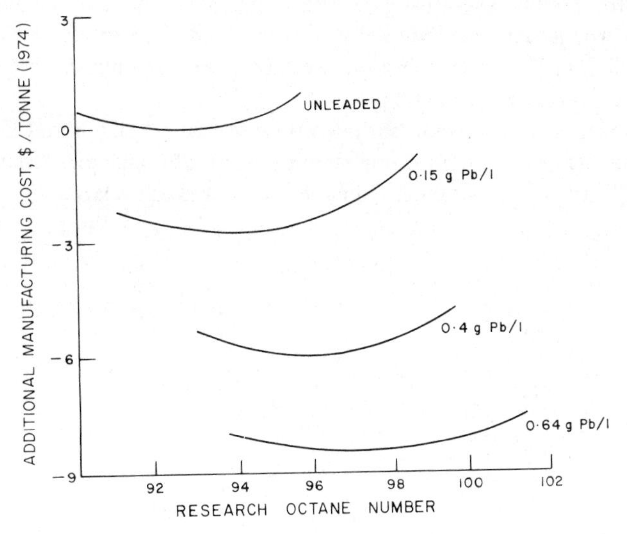

Figure 3.14 Additional cost of incremental octane numbers at various lead contents assuming that less-high-octane gasoline is required because of higher compression ratios[11]

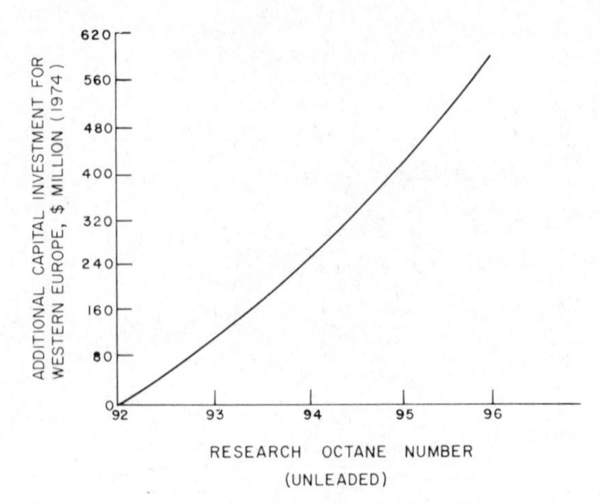

Figure 3.15 Additional capital investment for increased octane number in western Europe[11]

saving is not only on basic fuel cost but also on fuel tax: the higher the tax, the more will the motorist benefit from going to higher compression ratios (figure 3.16).

These calculations show, therefore, that there is no single optimum combination of compression ratio and fuel economy that offers an ideal solution from all points of view. Future engine designs will therefore almost certainly reflect a compromise between the desire of the motorist for maximum economy in terms of miles per gallon and the ability of the oil industry to justify the capital expenditure necessary to provide the required anti-knock quality. The investment required would be about $375 to save 1 ton of crude oil per year (1974 values), and this would not be economically attractive even at today's very high crude oil prices.

Under these circumstances, the only route that is really attractive to all parties is the development of high-compression-ratio high-efficiency engines that will run on low-octane-number gasolines, and as we have already seen

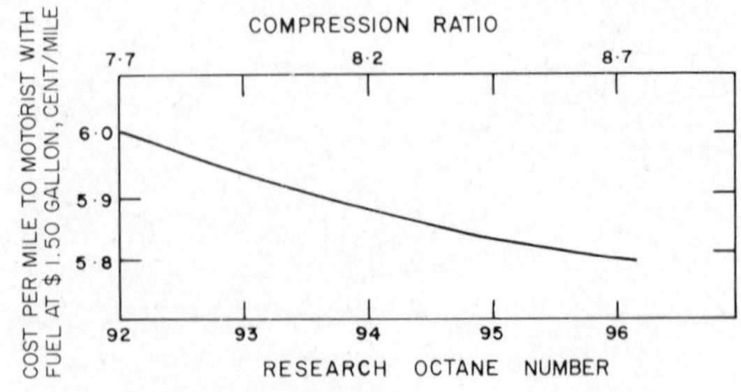

Figure 3.16 Effect of increasing octane number and compression ratio on the cost (1974) to the motorist, which includes tax

earlier in this chapter there are already indications that engine design may be moving in this direction.

One aspect of the economics of the use of gasoline that must also be mentioned is the question of the need for more than one grade. If, as has been assumed in the preceding paragraphs, the octane requirement of the car and the octane quality of the gasoline are perfectly matched, then there is no need for more than one grade. In practice, however, no matter how carefully the manufacturer builds his engines there will always be car-to-car variations in requirement, and the compression ratio will have to be set so that the highest-requirement cars are just satisfied by the best available gasolines.

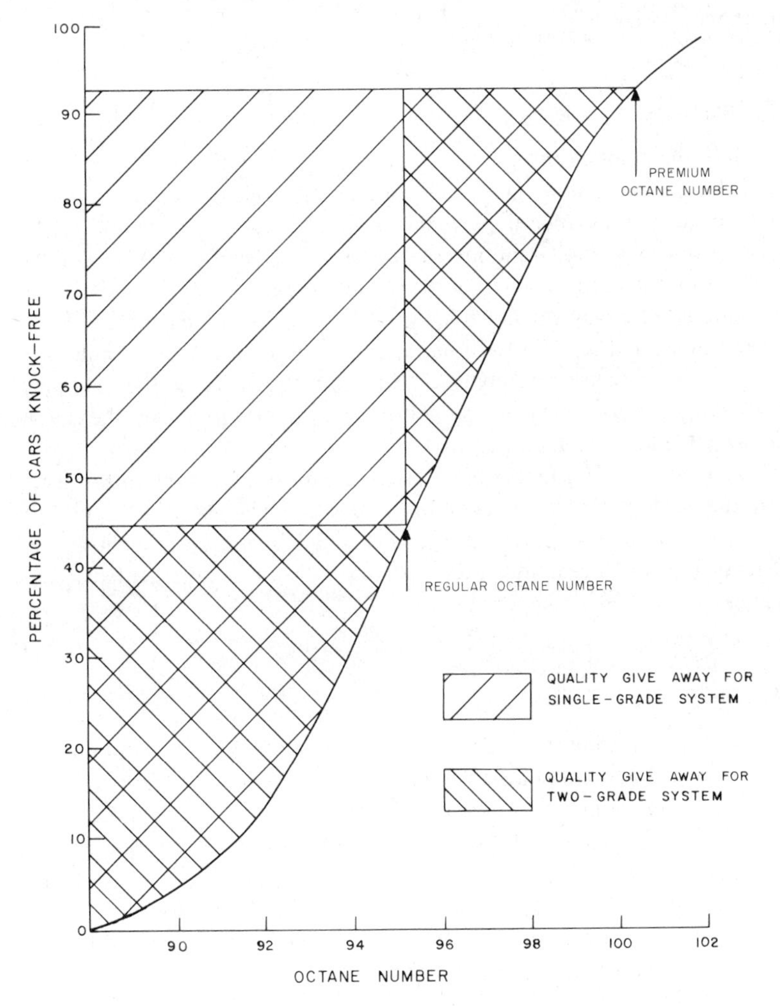

Figure 3.17 Diagram illustrating the saving in octane quality in going from a single-grade to a two-grade system

In a single-grade system this means that all the other lower-requirement cars will be forced to use a fuel of unnecessarily high quality. This is wasteful in terms of both cost and crude oil consumption and can be avoided by marketing gasolines of lower quality to satisfy the needs of the lower-requirement cars. In virtually all markets there are at least two grades available. Simple calculations have shown that, in a two-grade system, minimum manufacturing cost and optimum use of crude oil can be achieved when the second grade satisfies about half the cars in the market (figure 3.17).

Further, smaller reductions in cost and energy consumption can be obtained by using more than two grades or by the use of blending pumps, but against these savings must be set additional storage, handling, distribution and equipment costs.

3.5 Summary

For more than half a century the development of gasoline and the gasoline engine have gone hand in hand, and gasolines have been tailored to suit the needs of the cars in which they were to be used.

Decreasing availability and increasing cost of crude oil have placed greater emphasis on the need to conserve the world's energy resources, and economy will undoubtedly play a major role in the design of future engines.

One simple and direct way of improving economy (in terms of miles per gallon) is to increase the compression ratio, but the saving on the road may be offset by the additional energy consumed in the refinery to make the necessary high-octane gasolines.

Under these circumstances the most attractive route for the future seems to be the development of high-compression-ratio high-efficiency engines that will run on gasolines of current (or lower) octane quality; it may, however, be difficult to meet current and future exhaust emission requirements with such engines.

References

1. E. S. Corner and A. R. Cunningham. Value of high-octane-number unleaded gasolines in the US. Paper presented at *Am. Chem. Soc. Meet., Los Angeles, 28 March–2 April 1971*
2. D. F. Caris and E. E. Nelson. A new look at high-compression engines. *Soc. Automot. Eng. Prepr.*, No. 61A (1958)
3. O. Fersen. May's mpg Fireball. *Autocar* (6 March 1976).
4. British Technical Council. *The Efficient Use of Automotive Fuels: a Collection of Four Reports* (February 1976)
5. C. R. Major and A. C. Moore. Economics of anti-knock quality relating to motor gasoline. *J. Inst. Pet.*, **51** (1965) 215.

6. T. O. Wagner and L. W. Russum. Optimum octane number for unleaded gasoline. *Soc. Automot. Eng. Pap.*, No. 730552 (1973)
7. W. T. Tierney, E. M. Johnson and N. R. Crawford. Energy conservation optimization of the vehicle–fuel–refinery system. *Soc. Automot. Eng. Pap.*, No. 750673 (1975)
8. G. Salvi. Optimization of energy consumption in the refinery–fuel–vehicle system (in Italian). *Assoc. Tec. Automot.* (February 1976) 81
9. C. Marks and G. W. Niepoth. Car design for economy and emissions. *Soc. Automot. Eng. Pap.*, No. 750954 (1975).
10. J. J. Gumbleton, G. W. Niepoth and J. H. Currie. Compression ratio and fuel economy with emission constants. *Am. Pet. Inst. Prepr.*, No. 15-76 (May 1976)
11. H. van Gulick. Refineries and engines as a single technical system. *J. Automot. Eng.*, **6** (April 1975) 11

4 The Effect of the Physical Properties of Gasoline on Fuel Economy

B. D. CADDOCK

4.1 Introduction

The gasoline engine in its most widely used configuration invokes the use of a carburetter to meter liquid fuel into a moving airstream in response to the degree of throttling applied through the accelerator pedal. The rate at which this fuel is metered is intended to provide an appropriate compromise between requirements (a) to provide maximum power, (b) to provide maximum economy and (c) to comply with exhaust emission levels.

In any discussion of fuel economy it quickly becomes evident that a multiplicity of factors determine the economy that can be achieved in a particular situation with a single car. For example, the way in which the car is driven and the prevailing ambient conditions have a very large effect, as does the basic design of the vehicle, which takes into account variations in engine capacity, vehicle weight and resistance to tractive effort. In the present discussion the intention is to focus attention on an aspect of the subject that is closely linked to car design, driving pattern and ambient conditions, namely the effect of changes in the physical properties of the fuel.

Theoretical considerations indicate that changes in the following physical properties would be expected to influence fuel economy in appropriate circumstances.

(a) Net volumetric heating value.
(b) Specific gravity.
(c) Fuel volatility.
(d) Fuel viscosity.

The effects of these factors are reviewed in detail below. The effects of specific gravity and net volumetric heating value are interrelated and have therefore been taken together.

4.2 Specific Gravity and Net Volumetric Heating Value

Work in Shell laboratories has shown that the theoretical dependence of fuel economy on both heating value and specific gravity can be described by the equation

$$\text{mpg} = K(md + c) \tag{4.1}$$

where mpg is the fuel economy (in miles per gallon) of a fully warmed-up engine running at constant speed, K is a constant specific to a given car and speed, m and c are constants for the particular group of cars and gasolines concerned (i.e. US cars and fuels) and d is the specific gravity of the fuel.

This linear dependence of fuel economy on specific gravity is closely associated with the effect of the volumetric heating value, as shown by the equation

$$\text{mpg} = KQ_{\text{v}}\eta \tag{4.2}$$

where K is a constant, Q_{v} is the volumetric heating value of the fuel and η is the thermal efficiency.

The thermal efficiency of the engine process (see chapter 2) depends on the mixture strength F_{R} which is defined as F/F_{c}, where F is the fuel/air weight fraction and F_{c} is the stoichiometric fuel/air weight ratio. Thus, equation 4.2 can be rewritten as

$$\text{mpg} = f(Q_{\text{v}}, F_{\text{R}}) \tag{4.3}$$

An empirical linear relationship has been shown to exist between Q_{v} and d for gasolines such that

$$Q_{\text{v}} = md + c \tag{4.4}$$

This correlation is illustrated by the data plotted in figure 4.1. (From figure 4.1 it can also be seen that a similar correlation but with different parameters exists for pure hydrocarbons in the paraffin series, whilst pure aromatic hydrocarbons conform broadly to the pattern of behaviour of gasolines (see also chapter 2).)

This is an important relationship because the observed differences in the economy of fuels with different specific gravities are really due to differences in Q_{v}. The linear dependence of fuel economy on specific gravity stems from the secondary correlation of Q_{v} with specific gravity in accordance with equation 4.4.

From equation 4.3 it might be thought that the effect of fuel specific gravity on carburetter metering would have a direct influence on fuel consumption because of its effect on F_{R}, since simple carburetter hydraulics suggest that the metering of fuel into a moving airstream at constant engine speed is determined by the relations

$$F = K\text{d}^{1/2} \tag{4.5}$$

$$V = K'/\text{d}^{1/2} \tag{4.6}$$

where K and K' are functions of engine speed and nozzle dimension and V is the volume flow through the carburetter metering nozzle.

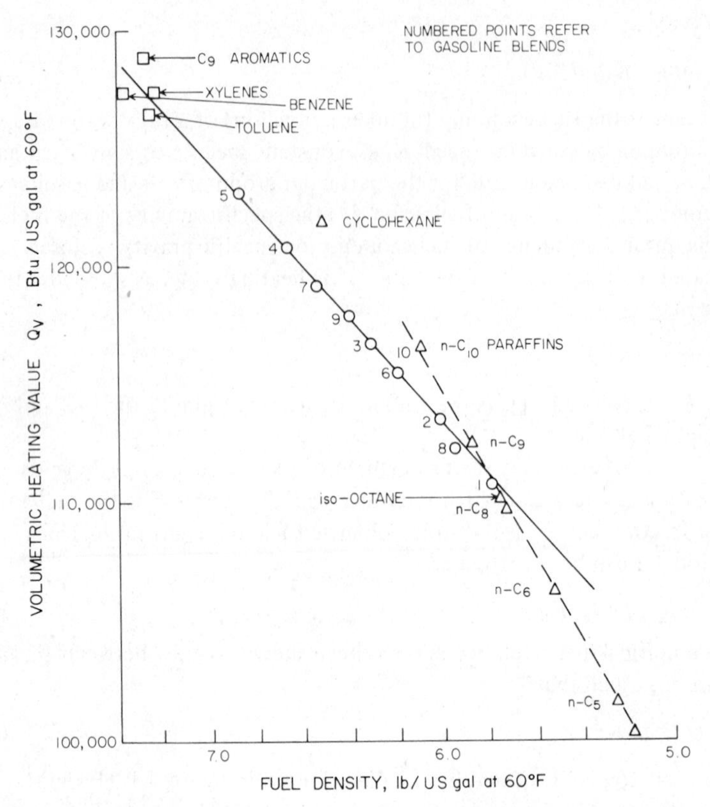

Figure 4.1 Gasoline heating value as a function of density (1 Btu/US gal = 0.278 kJ/l; 1 lb/US gal = 0.1198 kg/l)

It is fortuitous, however, that the effect of specific gravity on F is compensated by a similar effect on F_c, and in practice it is found that F_R is essentially independent of fuel specific gravity. Thus, the basic effect of fuel specific gravity on fuel economy is attributed to its influence on Q_v, as shown in figure 4.1. This is the theoretical basis of equation 4.1.

It would be expected therefore that a positive benefit in volumetric fuel economy would be observed as a consequence of increasing fuel specific gravity. An abundance of evidence is available to show that this is indeed the case in practice, and the dependence of fuel economy on specific gravity under fully warmed-up conditions has been demonstrated on many occasions. Although the precise relationship between economy and fuel specific gravity depends on the design features of an individual car an attempt has been made at Thornton Research Centre (TRC) to correlate results from bench engine tests with vehicle data. A number of laboratory results are plotted in figure 4.2. From these data it is concluded that an economy bonus of from 7 to 10% can be achieved by an increase in specific gravity of 0.1 under constant load conditions. Subsequent work has shown that under long-

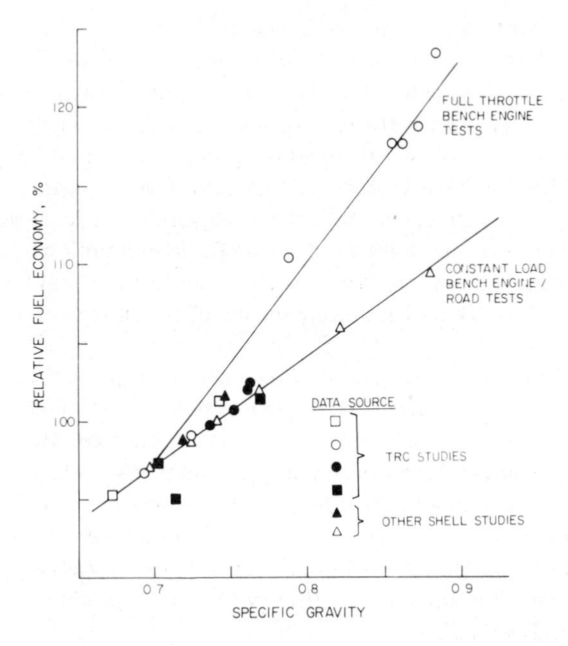

Figure 4.2 Correlation of specific gravity with relative fuel economy

trip road conditions benefits of from 7.5 to 8% were observed in practice with an increase in specific gravity of 0.1.

Since 1970 the impact of emission control legislation on car design has led to a general decrease in vehicle performance. These changes, however, although affecting the general level of fuel economy, appear not to have materially affected the response of US cars to changes in specific gravity[1].

Much of the knowledge available on the correlation of fuel specific gravity with fuel economy relates to richer-than-stoichiometric carburation regimes. Interest in achieving better fuel economy has stimulated interest in the use of leaner carburation with the associated development of improved mixture preparation systems and/or charge ignition systems. Although a similar dependence of fuel economy on specific gravity would be expected with such systems, little information is at present available to show the magnitude of the effect. A literature search for the period since 1970 did not produce any references (apart from reference 1) to published work on the measurement of the effects of specific gravity on fuel economy in modern vehicles.

4.3 Fuel Volatility

Specific gravity is the dominant fuel physical property in determining long-trip economy, provided that the fuel volatility is adequate to achieve efficient mixture preparation and distribution with a warmed-up engine. Increases in fuel specific gravity can be achieved by adding less-volatile components, but the benefit in fuel consumption achieved in this way would eventually peak and then start to decrease when the volatility reached a point where incomplete vaporization of fuel in the inlet system led to fuel maldistribution. With a cold engine in cold weather this is a problem that must be accepted with normal fuels, and so engines are fitted with chokes to enrich the mixture while the engine is warming up. It has been shown that under steady running conditions the use of the choke can increase fuel consumption by a factor of 2 or more, and it follows from this that a substantial benefit in fuel consumption should be possible by minimizing the use of the choke during the warm-up period.

The function of the choke is to maintain acceptable driveability whilst the car is warming up, i.e. to avoid effects such as hesitations, stumbles, stalls and backfires while the car is being driven following a cold start. Measurements on a chassis dynamometer have shown that driveability responds to fuel volatility and ambient temperature but that it is also very sensitive to the choke release pattern used during the test. There is, therefore, a trade-off, determined by the choke release pattern, between fuel economy and driveability, and this creates a difference in the way cars fitted with automatic chokes or manual chokes respond to changes in fuel volatility.

Keller and Byrne[2] demonstrated the effect of varying the choke release pattern and the fuel volatility on fuel economy during the warm-up period, and the results of their measurements, which are typical of similar measurements made at TRC and elsewhere, are shown in figure 4.3. The results in figure 4.3 suggest that with a fixed choke release pattern (i.e. an automatic choke) changes in fuel volatility have only a small effect on fuel economy during warm-up but that the use of a different choke release pattern has a substantial effect. This means that, for an automatic choke car with a fixed choke release pattern, changes in volatility would be expected to be largely reflected by changes in driveability rather than in fuel economy.

If a car is fitted with a manual choke, however, the driver has the option to release the choke to optimize either driveability or fuel economy. In some recent experiments with manual choke cars carried out at TRC cumulative fuel consumption and driveability measurements were made on two fuels with mid-points of 95 °C (368 K) (fuel B) and 109 °C (382 K) (fuel A), respectively. The results obtained on a Morris Marina 1800 are shown in figure 4.4 and again are typical for a manually choked car. Less choke was needed on fuel B to achieve satisfactory driveability, and a benefit in cumulative volumetric fuel consumption compared with fuel A is observed in the initial phase of the test. This benefit is gradually eroded by the effect of the higher specific gravity of the less volatile fuel as the engine warms up, and in this test a crossover was observed at a trip length of 11 km.

An important consequence of the relationship between driveability and fuel economy under cold-engine conditions concerns comparisons between

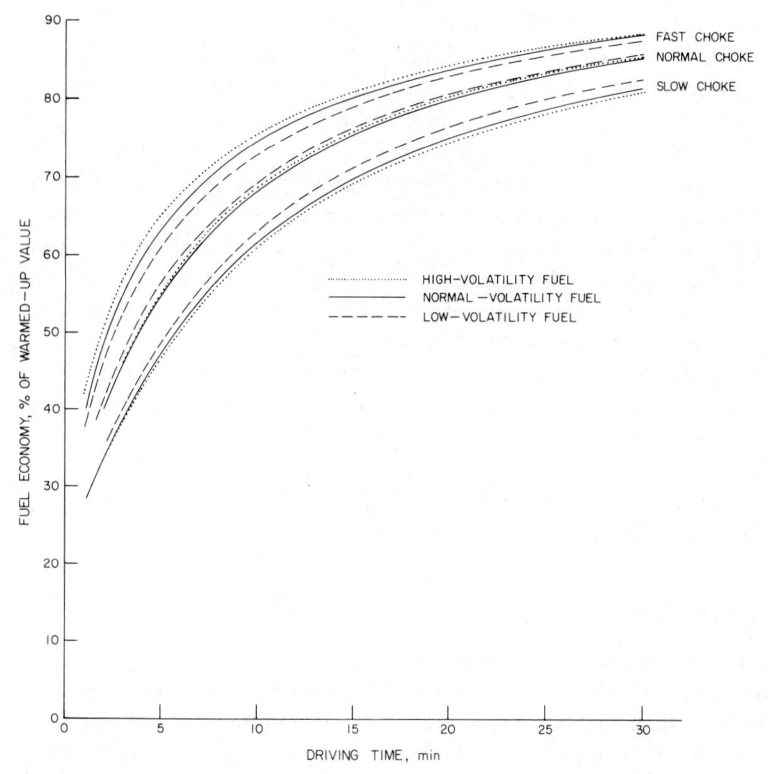

Figure 4.3 Warm-up fuel economy: effects of automatic choke setting and fuel volatility[2]

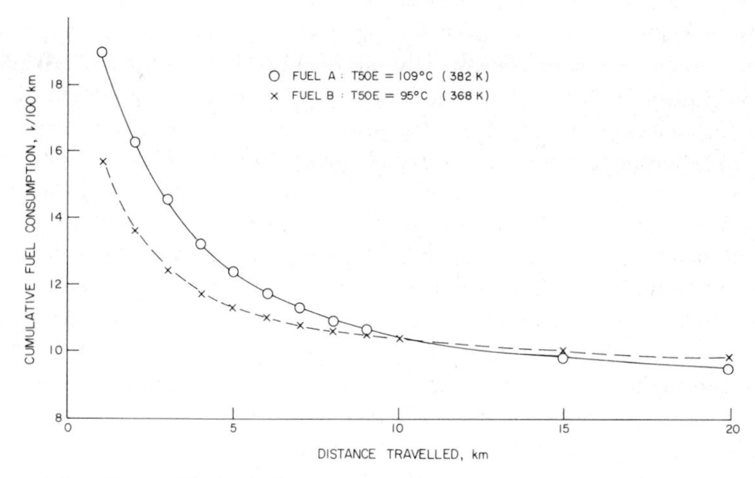

Figure 4.4 Effect of fuel volatility on warm-up fuel economy for manual choke Morris Marina 1800

the performances of fuels in relation to the effects of changes in fuel volatility. A rigorous comparison of fuel economies during warm-up should thus be made at constant driveability, whilst driveability performances should be compared at constant fuel consumption. It is evident that both aspects of performance need to be measured simultaneously if a comprehensive description of the effects of changes in fuel volatility on performance is to be obtained. This is an area where further data are needed, particularly with manual choke cars but also with automatic choke cars, and there is a need to define which volatility parameters are the most important in determining these effects. More field data are also needed in order to define how drivers with manual chokes actually use them.

It is concluded that there is a conflict in the requirements for better economy under short-trip conditions and under long-trip conditions. For short trips in cold weather a high-volatility fuel is beneficial for cars with manual chokes, provided the driver takes advantage of the better volatility to release the choke earlier. For long trips the best economy is always returned by the higher-gravity fuel.

4.4 Fuel Viscosity

In the past, fuel viscosity has not been considered to be of very great importance in carburation because gasolines all have very similar, very low viscosities, whilst the fuel orifices used in modern carburetters have in general a small length/diameter ratio. Additionally many carburetter designs feature air entrainment such that the fluid flow in the metering jet is non-laminar. Work at TRC has confirmed that the variation in viscosity for commercial gasolines is low. Experimental work at TRC and by Texaco[3] has shown that a gain in economy can be achieved by thickening gasolines with additives to induce gross increases in viscosity but that this gain is only achieved by causing the engine to operate under leaner conditions than the designer intended. Work in Shell laboratories has shown that the maximum difference expected from the differences in viscosity of normal gasolines would not be much greater than 1%. Carburetter metering has an obvious importance in determining fuel economy, and most texts on this topic use an empirical equation of the form of equations 4.5 and 4.6 to define the metered fuel/air ratio. In these equations the fuel is described only in terms of its specific gravity, and viscosity is neglected. It is an experimental fact, however, that metered mixture strength, particularly at idle, is sensitive to changes in ambient temperature. Increasing temperature leads to richer metering, and it is noteworthy that carburetter manufacturers have recently been experimenting with temperature-compensation devices to seek to control this effect. Precise metering is as critical in the context of emission control as it is in achieving maximum economy, and it is possible that the change in fuel viscosity with temperature, as well as the change in fuel specific gravity with temperature, will need more consideration in the future.

4.5 Discussion

In an era when energy conservation has been highlighted as a national objective in many countries, the achievement of even quite modest improvements in the overall fuel consumption of car populations can have a significant effect on crude oil importation. There are, for example, effects associated with small changes in fuel specific gravity and volatility which in relation to the activities of individual motorists are very difficult to demonstrate but which when integrated over a car population can become significant. A good deal of the knowledge available on these effects has been available for some years, but there are still some areas, especially concerning volatility effects, where data are somewhat limited.

In considering the optimum use of crude oil it is interesting that the overall effects of small changes in gasoline physical properties depend on which goal is being considered. From the motorist's point of view the main consideration is the possibility of a reduction in motoring costs. Since gasoline is sold on a volumetric basis the achievement of maximum miles per gallon is desirable and can be obtained by optimizing fuel volatility and specific gravity. The weight of evidence shows that under short-trip conditions good volatility is important for good fuel consumption (gravimetric and volumetric), whereas under long-trip conditions high specific gravity is needed for good volumetric fuel consumption. However, with notable exceptions, such as benzene, the specific gravity of gasoline blending components generally increases as the volatility decreases. A compromise must therefore be made in defining specification limits on volatility and specific gravity to control the relative levels of fuel economy and driveability given by the fuel. This does not conflict with the interest of refiners provided that the cost of doing this does not exceed the benefit realized by the motorist. This aspect of refinery economics needs to be more fully evaluated, although the differential cost of increasing gasoline volatility has been discussed briefly in a recent publication[4].

If, however, the goal is the national one of minimizing costs in relation to balance-of-payments problems, we should consider the refining and use of gasoline in the market as a single integrated system in order to estimate the minimum quantity of imported crude oil needed to achieve a given national mileage. The joint effects of fuel specific gravity and volatility on vehicle fuel economy then become only one of a number of contributory factors in this calculation. It is even possible that, in such an integrated approach, processing yield and energy consumption may dominate the situation to such an extent that a gasoline with low specific gravity and/or low volatility would give the best fuel economy from a given quantity of crude oil.

References

1. J. C. Ingamells. Fuel economy and cold-start driveability with some recent model cars. *Soc. Automot. Eng. Pap.*, No. 740522 (1974)
2. J. L. Keller and J. Byrne. What value front-end volatility? *Proc. Am. Pet. Inst., Sect. 3*, **46** (1966) 407
3. R. M. Reuter and G. W. Eckert. *US Patent*, No. 3,164,138 (5 January 1965)
4. H. van Gulick. Refineries and engines as a single technical system. *J. Automot. Eng.*, **6** (April 1975) 11

5 The Effect of Gasoline Additives on Fuel Economy

I. C. H. ROBINSON

5.1 Introduction

Crude petroleum consists essentially of hydrocarbons of various kinds boiling over a wide temperature range. In the early days of the spark ignition engine it was possible to use some of the light fractions as fuel without modification. Similarly it was possible to use selected heavy fractions as the crankcase lubricant. However, advances in engine and petroleum technology over the last several decades have led to the development of more efficient and much more powerful engines which need more complex fuels and lubricants for their operation. The performance of oil products can be improved to a substantial extent by selection and modification of the hydrocarbon structures, but there are practical and economic limitations to the performance levels that can be reached by hydrocarbon processing alone.

Today, the oil industry relies heavily on the use of additives to improve the performance of its products with respect to one (or more) important property[1]. We can define an additive as a material used at a low concentration either to improve the performance of a product with respect to some existing property or to confer on the product some entirely new property. Invariably, additives contain elements other than carbon and hydrogen; in the case of gasoline additives, compounds containing oxygen, nitrogen, lead, halogens and phosphorus have found wide application, and others containing boron, nickel, manganese, iron, etc., have also been used.

A convenient way of classifying gasoline additives is according to the applications for which they are intended, and the main classes are summarized in table 5.1. Minor applications such as anti-rust, anti-wear and the use of metal de-activators in long-life gasoline have not been included.

It is immediately apparent from this table that none of the major classes of additive in general use today has as its primary objective the reduction of fuel consumption. This is not really surprising when one considers that the fuel factor of greatest importance in the fully warmed-up engine is its calorific value. Gasoline hydrocarbons have high calorific values (about 43 MJ/kg), and

Table 5.1 Main classes of gasoline additive in use in 1975

Class	Date[a]	Typical compound	Typical concentration, ppmw	Comments
Anti-knock	1926	Tetraethyl lead	2000 (0.6 g Pb/l)	Requires the use of alkyl halides as lead scavengers
Anti-oxidant	1930	Alkyl phenol	100	Used in olefinic gasolines
Anti-icing	1952	Isopropyl alcohol	10000	Important in very volatile gasolines
Ignition control	1954	Tritolyl phosphate	200	Most effective in high-lead gasolines
Inlet system cleanliness				
First generation	1956	Alkyl amine phosphate	50	Effective in carburetter
Second generation	1971	Polyisobutene amine with a mineral oil carrier	500	Effective in whole of the inlet system

[a] Approximate date when the additive class became commercially important.

the possibility of substantially increasing the calorific value of a fuel by means of an additive used at a concentration of 1 or 2% appears remote.

What therefore is the relevance of the gasoline additive in the context of fuel economy? The answer is found in the efficiency with which the chemical energy of the fuel is converted into useful work. Heat engines are not noted for their efficiency, and the gasoline engine is not the most efficient heat engine. For example, in the US emissions driving cycle the average thermal efficiency of the gasoline engine has been found[2] to be about 10%.

There is clearly scope for improving this efficiency, and it is in this area that the fuel additive has a role to play. The role is a secondary one since vehicle design and operation are the dominant factors in fuel economy. Nevertheless, gasoline additives have an important part to play in (a) making it possible for basic engine design criteria to be met and (b) making it possible for design performance to be maintained over extended periods.

The contribution of the major classes of additive to fuel economy will be considered separately.

5.2 Anti-knock Additives

When Midgely and Boyd[3] discovered in 1922 that lead alkyls were very effective as anti-knock additives they were not seeking to improve fuel economy. Their problem was to avoid engine destruction arising from the mismatch between engine octane requirement and the low-octane quality of the fuel available at that time. For almost half a century, lead alkyls have been in virtually universal use in motor gasoline to raise its octane quality so that it can be used in higher compression ratio engines than would otherwise be the case. Although today we have the technology to make high-octane gasolines without lead, the use of lead alkyls remains the most cost-effective way of increasing octane quality[4]. It is legitimate, therefore, to examine the benefits which arise from their use.

Present-day lead contents are typically in the range 0.4–0.6 g Pb/l and result in increases in pool octane numbers of about from 5 to 6 units. The higher-octane gasoline permits increases in the compression ratio of about 1.5 units. The improved fuel consumption which results from the increased compression ratio amounts to about 6% on average (see chapter 3). Thus we can regard anti-knock additives as contributing to improved fuel consumption provided that engine requirement and fuel octane level are matched. In an engine which is knock-free on a given fuel the addition of an anti-knock additive would confer no benefit. Thus, an anti-knock additive can contribute to improved fuel economy if it improves the matching of fuel octane number to engine requirement. At today's compression ratios the contribution of an anti-knock additive is about 1% improvement in fuel economy for each octane number increase.

5.3 Anti-oxidants

The use of anti-oxidants can be said to have helped to improve fuel economy over the years but in a less tangible way than anti-knock additives. Anti-oxidants were introduced in order to prevent oxidation in storage of cracked gasoline components. Historically, thermal and catalytic cracking processes were introduced to increase the yield of gasoline per barrel of crude oil. The cracked components (olefinic hydrocarbons) were found to have higher octane numbers than straight-run paraffinic gasoline had, and this allowed increased compression ratios (with concomitant gains in fuel economy). However, the olefinic components oxidized easily, giving rise to pro-knock peroxides and gum formation in the engine unless corrective action was taken.

Thus, the contribution of anti-oxidants to fuel economy can be seen as one which permits the inclusion of less-stable but higher-octane components in motor gasoline. The additive role is a limited one, however, as gasolines made by catalytic reforming seldom need the addition of anti-oxidants. It is therefore not possible to quantify the contribution to fuel economy arising from the use of anti-oxidants. Their major benefit is in improving the flexibility of gasoline manufacture.

5.4 Anti-icing Additives

Evaporation of gasoline in the carburetter of an engine results in a drop in temperature of the fuel–air mixture and the carburetter. If the air is humid and the temperature drop is sufficient to produce sub-zero temperatures, ice will form in the carburetter. The worst conditions are encountered at air temperatures of around 5 °C and at relative humidities greater than 90%[5].

Apart from the impairment in driveability which results from carburetter icing, there may also be a fuel consumption penalty. The extent to which fuel consumption will be affected by carburetter icing will depend on many factors which include (a) carburetter design, (b) inlet air temperature, (c) inlet air humidity and (d) fuel volatility.

Under conditions conducive to carburetter icing the adverse effect of icing on economy can readily be demonstrated (figure 5.1). The corrective effect of an anti-icing additive (dipropylene glycol (DPG) in this experiment) is very clear. Road tests at the Thornton Research Centre (TRC) carried out under severe icing conditions have shown improvements in steady-state fuel consumption of up to 20%.

Thus, the driver of a car which is sensitive to carburetter icing and who lives in an area with the right climatic conditions will undoubtedly benefit from the use of anti-icing additive in his fuel. However, the overall contribution of anti-icing additives is probably very small. Sensitive cars are in a rapidly decreasing minority in the market-place as a result of engine design changes which have taken place over the last several years. Most new cars

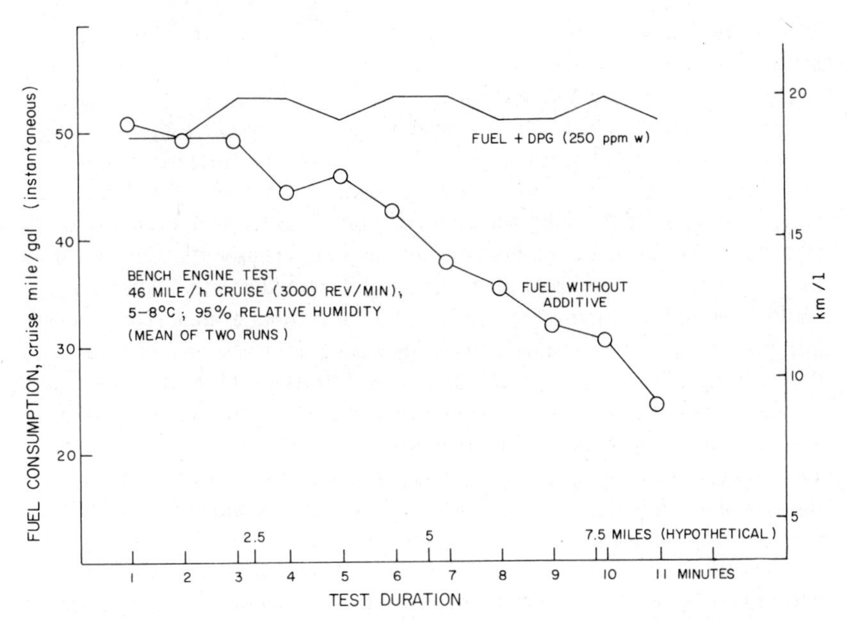

Figure 5.1 Fuel consumption in a Ford 105E bench engine: icing conditions

today are fitted with a system for pre-heating the air supplied to the carburetter. This is done largely to maintain good driveability with the lean mixtures needed to reduce exhaust emissions but also has the advantage of virtually eliminating carburetter icing.

We conclude therefore that, although there are individual motorists who are undoubtedly achieving better fuel consumption as a result of anti-icing additives, they are part of a very small and dwindling minority. The effects on fuel economy although directionally favourable are probably not significant in market volume terms.

5.5 Ignition Control Additives

Over a period of time, deposits slowly accumulate in the combustion chamber of the gasoline engine with certain adverse effects on engine performance. Two of the effects are susceptible to treatment with phosphorus-containing fuel additives, which have become known as ignition control additives. Both problems have as their root cause the deposition of lead salts after combustion of lead-containing fuels.

Lead alkyls used as anti-knock additives are invariably used with alkyl halide (dichloroethane and/or dibromoethane) scavengers whose purpose is to convert the lead into lead halides which are readily exhausted from the combustion chambers. Halogen compounds were chosen for this role because lead halides are relatively volatile at combustion chamber temperatures and little of the lead remains in the chamber. However, the little that does remain can give rise to operational problems.

Since the electrical resistance of lead halides falls rapidly with increasing temperature, lead chlorides and bromides depositing on the ceramic insulators of spark plugs can give rise to spark plug misfire by providing an alternative path to earth. Lead phosphate has a much higher electrical resistance than lead chloride or bromide compounds have, and in the early 1950s it was found that the addition of aryl phosphates to gasoline would greatly alleviate and could often prevent spark plug misfire[6], by converting residual lead halides into phosphate. Aryl phosphates have since been used as supplements to the lead scavenger but at much lower molar concentrations. (Lead scavengers are normally used at one to one-and-a-half times the theoretical concentration T needed to convert the lead in the fuel into lead halide, whereas phosphorus additives are normally used at $0.1-0.3T$.) The intention is that the halide should still exercise its role as a scavenger but that any remaining lead should preferentially be in the form of the phosphate.

The other benefit that accrues from having lead deposits present as phosphates rather than as halides is the reduced tendency to surface ignition. Lead halides catalyse the ignition of carbonaceous deposits, whereas lead phosphates are relatively inert in this respect. Surface ignition was a serious problem in the US for some years, although much less so in Europe, and manifested itself as abnormal combustion noise during acceleration after a period of low-duty operation. It was caused by spontaneous ignition of the fuel–air charge by glowing deposits; the noise resulted from the very high rates of pressure rise. Mostly the problem was confined to one of excessive combustion noise, but in some circumstances surface ignition could lead the engine into pre-ignition and engine failure. Phosphorus additives greatly alleviated the situation but did not prevent it completely in a severe engine[7].

The benefits with respect to fuel consumption which accrue from the use of phosphorus additives have been studied. Particularly in the case of spark plug fouling, it is possible to demonstrate that spark-plug misfire increases fuel consumption and that the use of phosphorus additives can restore fuel consumption to that achieved with clean plugs (figure 5.2). However, the use of phosphorus additives today is on the decline. Of the many reasons for the decline the major ones are as follows.

(1) Gasoline lead contents are being reduced.

(2) Phosphorus has an adverse effect on the life of exhaust catalysts.

(3) Newer spark plug designs are less prone to fouling.

(4) Present-day engines are much stiffer and exhibit much less abnormal combustion noise.

(5) Both spark plug misfire and surface ignition are transient phenomena which can usually be cleared by a short period of full-throttle operation.

It would appear, therefore, that, although there are several benefits from

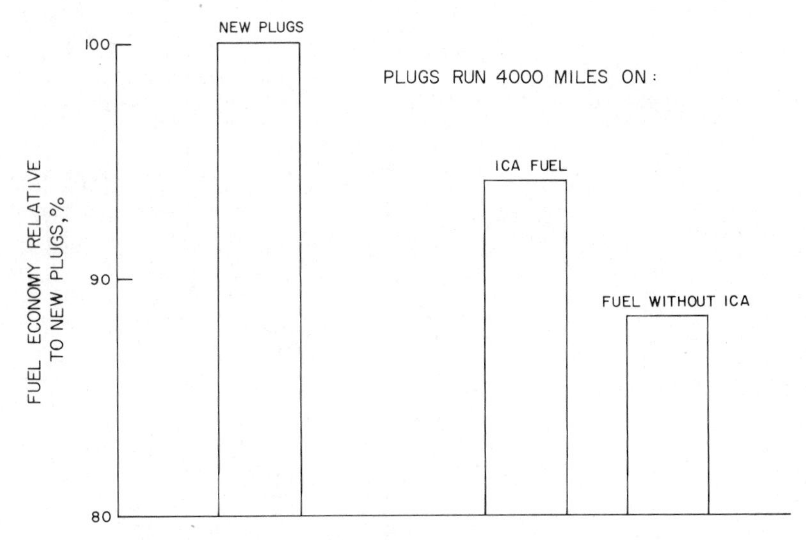

Figure 5.2 Improvement in gasoline economy with ignition control additive (ICA)[7]

the use of phosphorus additives, the need for such additives is decreasing. Exhaust catalyst life is greatly enhanced by the absence of lead, halogens and phosphorus from motor gasoline. We therefore conclude that, while phosphorus additives had a minor role to play with respect to fuel consumption in past years, such additives will slowly disappear from gasoline.

5.6 Carburetter and Inlet System Cleanliness Additives

As combustion chamber deposits accumulate over a period of time, so also do deposits in the carburetters and inlet systems of gasoline engines. These deposits can interfere with the correct functioning of the engine, and their presence is best avoided[8,9].

The most recent major development in the gasoline additive field has been the emergence of a second and much more effective generation of inlet system cleanliness additives. When carburetter detergents were first introduced in the US in the 1950s, they made relatively little impact. Their use resulted in cleaner carburetters and perhaps minor benefits with respect to carburetter icing and fuel tank rusting[10], but their value to the user was not clearly substantiated. However, as attention was focussed on the passenger car as a source of air pollution, it became increasingly clear that inlet system cleanliness additives could play some part in the drive for cleaner air[11].

A new car has its carburetter designed and adjusted to provide the optimum air/fuel ratio intended by the car manufacturer. However, over a period of time deposits build up in the carburetter, and the air/fuel ratio departs from the design condition. In general, the effect of carburetter

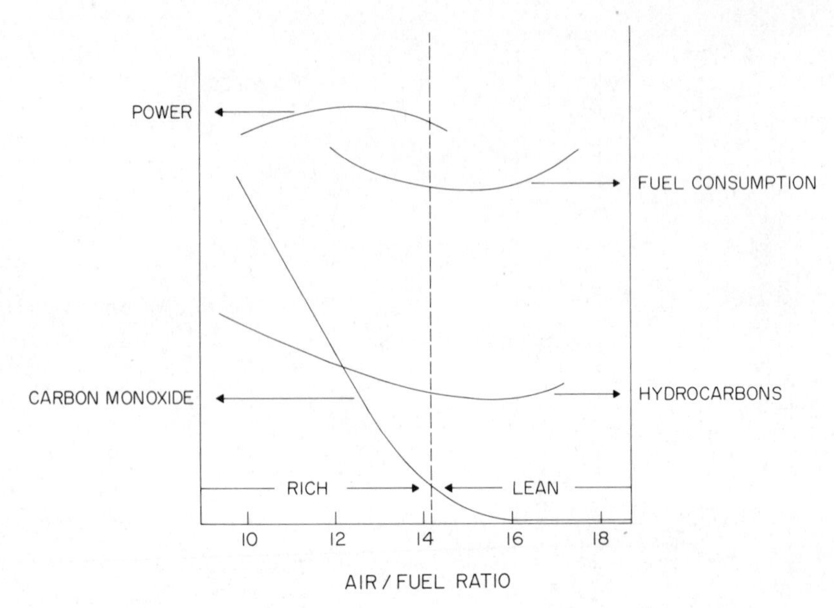

Figure 5.3 Influence of air/fuel ratio on power, fuel consumption and emissions

deposits is to restrict the air flow with the result that the mixture becomes richer as deposits accumulate[8]. As a consequence, exhaust emissions of carbon monoxide and hydrocarbons are increased and so is the consumption of fuel (figure 5.3). It follows, therefore, that control of deposits by means of fuel additives offers benefits not only to the user but also to the community. This then is the background to the development of a new class of cleanliness additive which not only prevents deposit formation but also removes pre-formed deposits[12]. In addition, a second-generation additive controls inlet valve deposit formation, an area in which the carburetter detergent is ineffective[11].

Sensitive areas for deposit formation are carburetter throttle bodies, air bushes in carburetter idle and progression systems, positive crankcase ventilation (PCV) valves and exhaust gas recirculation (EGR) systems. Deposit build-up in any of these areas can have a substantial effect on air/fuel ratio and hence of fuel economy and emissions.

The adverse effects of carburetter deposits can readily be demonstrated in laboratory bench engine tests in which crankcase blowby gases and exhaust gases are recirculated to the carburetter to accelerate deposit build-up. For example, in a standard carburetter cleanliness test procedure using a Ford Escort engine at TRC, increases in fuel consumption and exhaust emissions can readily be correlated with the build-up of throttle body deposits (figure 5.4).

The effect of air brush deposits can be demonstrated in a Fiat 600 bench

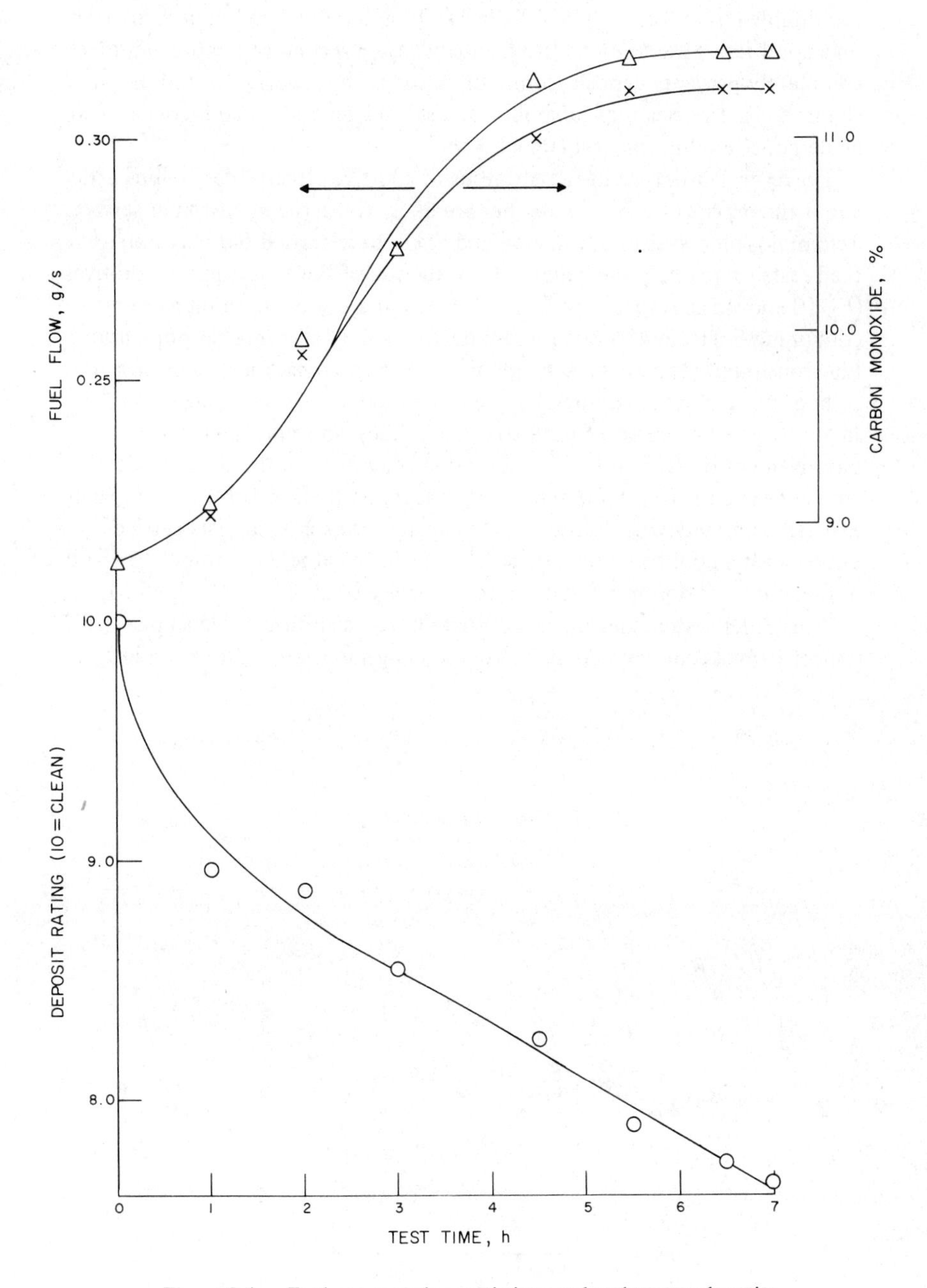

Figure 5.4 Fuel consumption, emissions and carburetter deposits

test. Using an accelerated deposition procedure, exhaust CO concentration can double (from 3 to 6%) in as little as 4 h on base fuel and in perhaps 10 h on a good first-generation additive. In contrast, a second-generation additive completely prevents deposit formation in the air bush for at least 40 h (figure 5.5). The clean-up capability of a second-generation additive can also be demonstrated in this test (figure 5.5).

However, laboratory demonstrations of additive effectiveness tell us little about the extent of the benefits that are likely to accrue in customer service. Two major oil companies, Chevron and Shell, have carried out very extensive road tests to quantify the benefits from the use of their proprietary additives (F-310 and ASD, respectively). After 2000 miles of operation on gasoline containing F-310 in 450 cars representative of the California car population, Chevron found reductions of 13.9% in exhaust hydrocarbons and 11.6% in carbon monoxide as measured by the seven-mode emissions cycle (hot)[14]. In another test programme in thirteen cars, Chevron found an average improvement of 7.7% in fuel consumption (measured in the seven-mode emissions cycle) after 2000 miles of operation on their additive fuel[12]. Shell experience mainly with European and Japanese cars in road tests carried out in several countries indicates reductions in CO at idle of around 15% and in fuel consumption in normal customer service of about 4% (figure 5.6).

Thus, inlet system cleanliness additives have a significant role to play with respect to fuel economy. An effective second-generation additive can largely

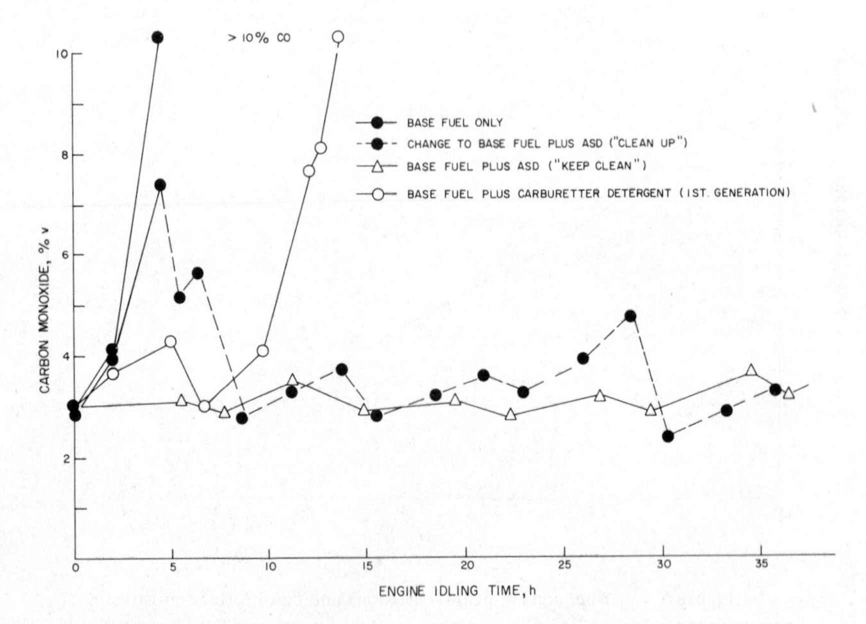

Figure 5.5 The effect of carburetter additive detergents on CO emission at engine idle – Fiat 600; Solex carburetter

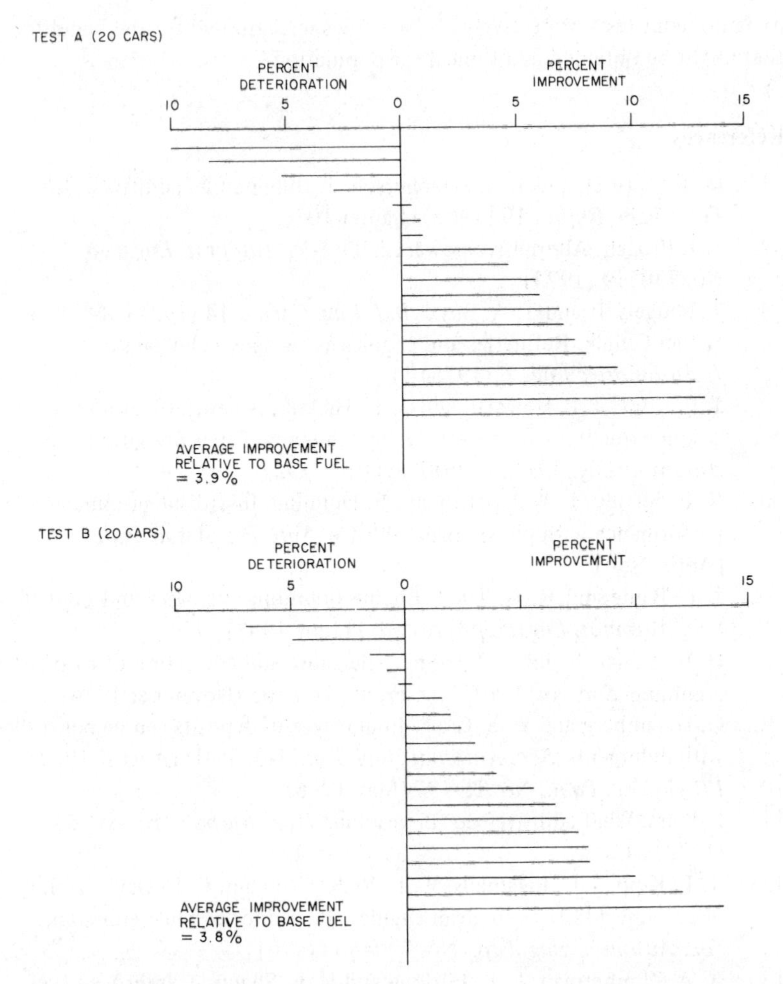

Figure 5.6 Individual improvements in consumption (mileage) on ASD fuel relative to base fuel after 6000 km (4000 miles) operation

prevent deposit formation in the carburetter and inlet system and thus preserve the air/fuel ratio settings intended by the engine manufacturer.

5.7 Other Additives

The use of gasoline additives to improve driveability was proposed by Exxon in 1971. The additive HTA was claimed to render the inlet manifold walls oleophobic, thus reducing wall wetting by gasoline and improving mixture distribution. In cars with mixture distribution problems Exxon reported improvements in fuel consumption of 3.3% and 4.5% in road and chassis dynamometer tests respectively[13]. No values were quoted for the benefits that might be obtained in a typical car population.

References

1. M. R. Barusch and J. H. Macpherson, Jr. Engine fuel additives. *Adv. Pet. Chem. Refin.*, **10** (1965) chapter 10
2. J. J. Brogan. Alternative powerplants. *Soc. Automot. Eng. Pap.*, No. 730519 (1973)
3. T. Midgely Jr and T. A. Boyd. *Ind. Eng. Chem.*, **14** (1922) 589, 894
4. H. van Gulick. Refineries and engines as a single technical system. *J. Automotive Eng.*, **6** (1975) 11
5. J. F. Kunc, J. P. Howarth and J. F. Hickok. A new look at motor gasoline quality – carburetter icing tendency. Paper presented at *Soc. Automot. Eng. Meet., Detroit, January, 1951*
6. R. E. Jeffrey, L. W. Griffith and E. Dunning. Improved gasoline performance with phosphorus additive. *Natl. Pet. Assoc. Pap.* (April 1954)
7. J. L. Bame and R. G. Tuell. Engine pounding, its cases and control. *Soc. Automot. Eng. Prepr.*, No. 61G (June 1958)
8. H. W. Sigworth and J. Q. Payne. The cause and correction of carburetter gumming. *Soc. Automot. Eng. Prepr.*, No. 405 (November 1954)
9. G. H. Amberg and W. S. Craig. Intake system deposits can be controlled with detergents. *Soc. Automot. Eng. Pap.*, No. 554D (August 1962)
10. *Ethyl Corp. Publ.*, No. TS-197 (May 1968)
11. P. Polss. What additives do for gasoline. *Hydrocarbon Process.*, **52** (1973) 61
12. K. L. Kipp, J. C. Ingamells, W. L. Richardson and C. E. Davis. Ability of gasoline additives to clean engines and reduce exhaust emissions. *Soc. Automot. Eng. Pap.*, No. 700456 (1970)
13. A. A. Zimmerman, L. E. Furlong and H. F. Shannon. Improved fuel distribution – a new role for gasoline additives. *Soc. Automot. Eng. Pap.*, No. 720082 (1972)
14. Chevron presentation, Los Angeles (August 1970)

6 The Effect of Mixture Preparation on Fuel Economy

G. A. HARROW

6.1 Introduction

The quality of the fuel–air mixture supplied to a gasoline engine affects its fuel economy in many ways, some of which are subtle and are not fully understood.

The main reason for this is that mixture quality plays an 'enabling' role, i.e. perfect mixture preparation would enable a given engine structure to function at its maximum thermal efficiency, whereas poor mixture preparation would not. Equally, if perfect mixtures could be achieved, engine designs could be tailored to exploit the benefits they confer.

There is therefore no clear relationship between mixture quality and fuel economy because it is almost impossible to isolate the mixture preparation system from the engine design as a whole. It is particularly difficult to be sure that benefits found for some particular type of mixture preparation system are present under all operating conditions and are uniquely due to the better mixture quality.

Another complication is that traditionally, at full throttle, part of the latent heat of vaporization of the fuel is used to chill the mixture entering the engine and thus to increase the maximum power. In the absence of a perfect mixture preparation system which will provide both a chilled mixture and a fully atomized fuel, engine inlet manifolds have to cope with mixtures of liquid and vapour. A consequence of this need to combine several functions is that any engine mixture preparation system is a package which represents the manufacturer's best compromise between the conflicting requirements of power, emissions and economy. The result is invariably sensitive to minor changes in the carburation and inlet system geometry, which can give an improvement for one type of engine condition at the expense of performance in another. Attempts to identify, by direct experiment, the consequence of improved mixture preparation alone are therefore full of pitfalls, and the interpretation of the results of such experiments must be approached with care.

Exploitable fuel economy benefits will, in fact, only be obtained via the mixture preparation route if the mixture quality can be improved for all engine conditions and if the carburetter or fuel injection system can be retuned to make the best use of these improved mixtures. The improved fuel economy will then be due to the more economical mixture strength regimes made possible by the better mixture preparation system.

In this chapter we start from the fact that all engines can be induced to operate in regimes of high thermal efficiency, and we try to work back to the type of mixture quality appropriate for each operating condition. By this means many problems of interpretation can be avoided, and a clearer picture can be drawn of the demands placed by engines on their mixture preparation systems. The purpose of the chapter is to survey some of the available information on the way in which fuel is consumed by gasoline engines on the road and to infer from this the most important effects of mixture quality on fuel economy.

6.2 Mixture Quality in Current Carburetted Engines

We would expect that when a fuel passes through the carburetter of an engine it will emerge as a homogeneous mixture with air which will be distributed uniformly to all the cylinders. In practice, however, this is not so, and, as shown by Skripkin et al.[1] and by many other workers[2–9], there is little turbulent mixing of the fuel with the air in the inlet system. Increased concentrations of fuel resulting from the deflection effect of the throttle plate or the corresponding position of the fuel nozzle are maintained right into the inlet manifold branch. The presence of a mixing device after the carburetter improves the homogeneity of the mixture and permits a better fuel distribution between the cylinders, but it imposes a power loss on the engine. Normally the fuel entering the cylinders of an engine does so by three routes: as a vapour, as a mist of droplets created in the carburetter and as a film of liquid attached to the inlet manifold wall. Plate 6.1 is a close-up photograph of the liquid fuel film in a glass manifold fitted to a conventional four-cylinder engine. In the case of a metal manifold the fuel film would be thinner because the heat conducted from the hot spot and the cylinder head would help the fuel to evaporate. Nevertheless this sort of liquid film can occur in any inlet manifold if the conditions are right. As the engine speed and operating conditions alter, so do the film thickness and the distribution of liquid fuel between the cylinders. The amount of fuel 'held up' on the manifold wall is an important property of the engine, because when the throttle is suddenly closed during a deceleration most of it is flashed off and passed unburnt through the engine as an overrich mixture. The manifold will thus dry out during overrun, and, when the throttle is subsequently snapped open, a liquid layer will have to be re-established before all the cylinders can be supplied with equal amounts of fuel. Whilst this is happening, the weaker cylinders may misfire, and the engine will stumble.

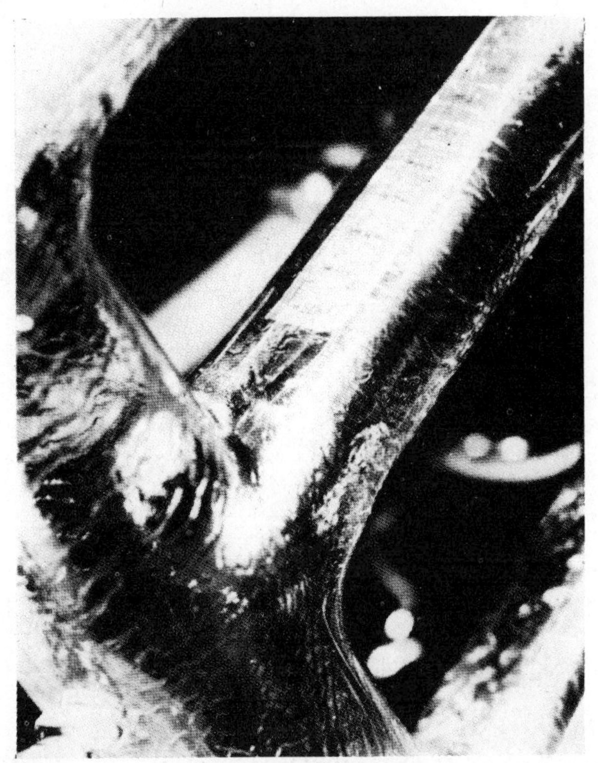

Plate 6.1 Close-up of liquid fuel behaviour in a glass inlet manifold of a multi-cylinder gasoline engine at full throttle

This sequence of events can be clearly identified in the composite trace of air/fuel ratio and inlet manifold pressure shown[10] in figure 6.1. Starting at the right-hand side of the figure, the engine is idling at 1100 rev/min with the throttle closed, and the air/fuel ratio is approximately 11.2/1. At point A the throttle is suddenly opened, and the engine is allowed to accelerate. The mixture momentarily weakens to 14/1 air/fuel ratio at B and then becomes richer as the engine speeds up to about 1900 rev/min. At C the throttle is suddenly closed, and the air/fuel ratio moves from 11.9 to 5.0 as the inlet manifold pressure falls. Finally the mixture weakens as the engine is allowed to slow down from about 2040 rev/min to 1550 rev/min at D.

In varying degrees, this type of mixture strength excursion occurs in all carburetted engines, fixed-jet carburetters exhibiting large excursions and variable-jet carburetters smaller ones. Each cylinder of the engine is subjected to this sort of mixture strength variation with the weaker cylinders having a larger mixture strength excursion corresponding to point B than the richer ones. If the weak-mixture limit is exceeded, then that cylinder will misfire. We can appreciate therefore that in any engine the presence of the liquid film gives rise to two sorts of problem, one in which there is a mixture strength difference between adjacent cylinders and the second in which, as a

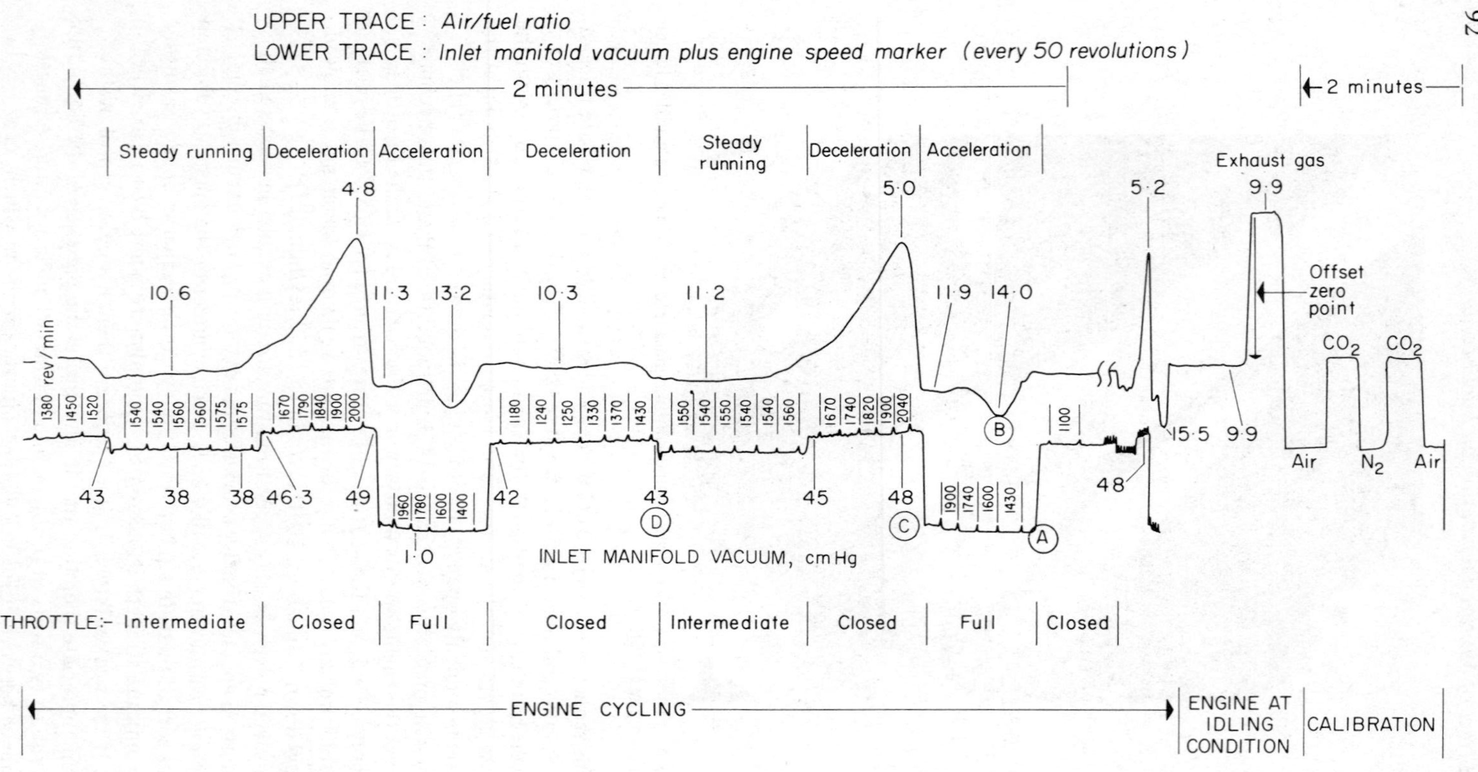

Figure 6.1 Engine cycling test (1 cm Hg manifold vacuum = 1.333 kN/m²)

consequence of the re-establishment of a liquid film during acceleration, some cylinders may malfunction during transient conditions of engine operation.

6.3 Mixture Maldistribution between Cylinders

There are two principal reasons why liquid fuel is present in the inlet manifold of an engine: one is that insufficient heat has been supplied to vaporize the fuel, and the second is that there is insufficient mixing of the fuel and air to yield a homogeneous mixture.

The engine manufacturer tries to ensure uniform distribution of fuel between the cylinders by providing a hot spot to transmit exhaust heat to the mixture and by drawing some of the inlet air to the engine from around the exhaust manifold. The inlet manifold is also designed to permit a certain amount of liquid fuel to flow freely to the cylinders. These procedures are not normally very effective, and almost invariably some fuel maldistribution does occur. In table 6.1 the measured air/fuel ratio spreads amongst the cylinders are given for some four-cylinder car engines running at road load and at full throttle. It is apparent that most attention has been paid to the road load condition, the 2.0 l and 1.9 l cars being particularly good. At full throttle, apart from the 1.6 l car, the extent of fuel maldistribution varies with engine speed, and the full-throttle condition has obviously been sacrificed to ensure good mixture distribution at road load. Transient operation of an engine soon shows up the problems, and the composite trace of CO emission from individual cylinders of a 1.5 l engine[11] shown in figure 6.2 indicates that, even though the steady-state mixture distribution may be good,

Table 6.1 Typical spread of air/fuel ratios between cylinders for standard road vehicles

Engine mode[a]	Maximum air/fuel ratio range between cylinders						
	1.3 l, make A	1.8 l, make A	1.5 l, make B	1.6 l, make B	2.0 l, make C	1.9 l, make D	2.3 l, make E
Idle		0.4	1.7	0.3	0.3	1.1	0.4
30 mile/h RL	0.4	0.4	0.5	0.8	0.2	0.4	0.4
50 mile/h RL	0.7	0.9	0.8	1.5	0.7	0.5	0.6
70 mile/h RL	2.8	1.1	1.6	0.8	0.3	0	1.5
30 mile/h FT	2.2	1.5	2.5	1.0	1.6	1.8	3.5
50 mile/h FT	1.5	3.2	1.9	1.0	2.2	2.9	0.9
70 mile/h FT	2.0	2.5	2.2	1.1	1.6	3.4	1.6

[a]RL, road load; FT, full throttle.
1 mile/h = 1.609 km/h.

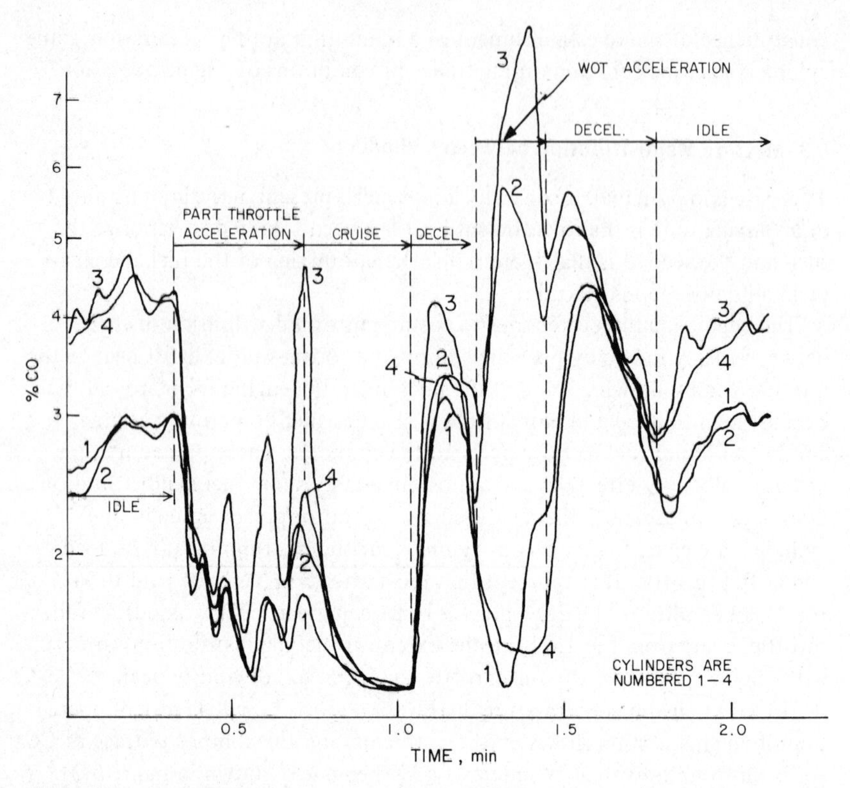

Figure 6.2 Individual cylinder exhaust CO levels, illustrating excellent distribution in steady-state conditions but poor in transient conditions

once the engine is accelerated, one of the cylinders (No. 3) becomes much richer than the others.

Motor gasoline is a mixture of hydrocarbons most of which boil over the temperature range 30–200 °C. Also present are some compounds which boil at low temperatures (e.g. butane, b.p., –0.5 °C) and very small quantities of material which boil at higher temperatures (e.g. tetradecane b.p., 253.6 °C). It is not necessary to heat the mixture to its final boiling point in order to achieve full vaporization. If we know the air/fuel ratio of the mixture, the fuel composition and the inlet manifold pressure, then we can calculate the minimum temperature at which evaporation should theoretically be virtually complete. We can call this the dew-point temperature. Provided the mixture temperature is well above the dew-point, then all the fuel should be vaporized, and there should be little fuel maldistribution between the cylinders. Typical dew-point curves for a premium gasoline are given in figure 6.3. Figure 6.4 shows, for a modern engine, the relationship between the mean air/fuel ratio, the extent of mixture superheating above the dew-point and the extent of

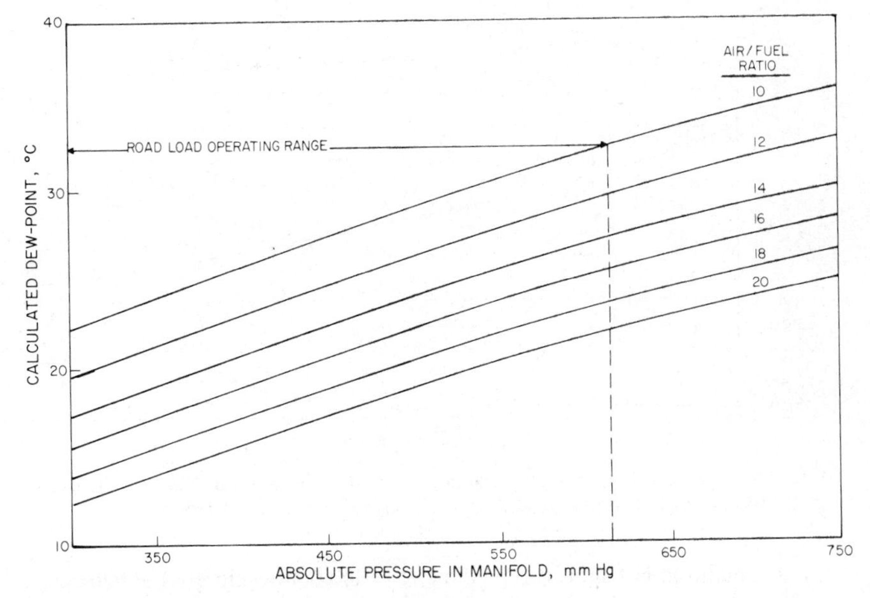

Figure 6.3 Calculated dew-points for a premium gasoline at different values of the absolute pressure (1 mm Hg = 0.133 kN/m^2)

mixture maldistribution for various speeds at road load. It can be seen that up to 50 mile/h (80.5 km/h) the mixture temperature is well above the dew-point and that there is no significant fuel maldistribution. For all practical purposes we can assume that the mixture preparation is satisfactory. Above 50 mile/h (80.5 km/h), however, the mixture temperature approaches the dew-point, and the extent of fuel maldistribution between the cylinders rises to about 4 units of air/fuel ratio. If fuel maldistribution between the cylinders is substantial during steady-state operations, then it may become very much worse during transients on the road. The mixture strength in some cylinders may exceed the weak-mixture limit and the engine will then misfire.

Enriching the mixture when engine malfunctions occur is a common way of reducing the effect of faulty mixture preparation. It is apparent when the full-throttle condition is considered, as in the case shown in figure 6.5. Here, for the same engine, the mixture strength, the extent of charge superheating and the extent of maldistribution are plotted as a function of road speed at full throttle. The mixture temperature is always close to the dew-point, and the mixture strength is quite rich (about 13/1 air/fuel ratio). The low charge temperature has been deliberately sought by the manufacturer so as to maximize the full-throttle power, but as a consequence much liquid is present in the manifold and the fuel distribution amongst the cylinders is poor. By calculating the amount of heat needed to vaporize the measured fuel flow to the engine and by comparing this with the amount of heat actually supplied

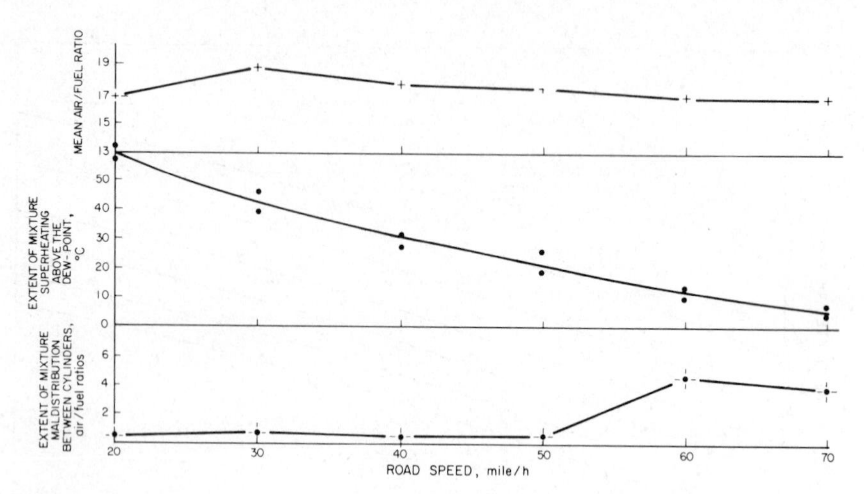

Figure 6.4 Variation of mean air/fuel ratio: extent of mixture superheating and extent of mixture maldistribution at road load (1.7 l engine) (1 mile/h = 1.609 km/h)

by the manifold hot spot, it is possible to estimate how closely the two agree, and this is done in figure 6.6. It can be seen that for the road-load condition enough heat is supplied by the hot spot, and so the problem there is simply insufficient mixing. At full throttle not enough heat is supplied to the mixture, so that the problem there is one of both heating and mixing. The gap between the full-throttle heat supply and the demand curve shown in figure

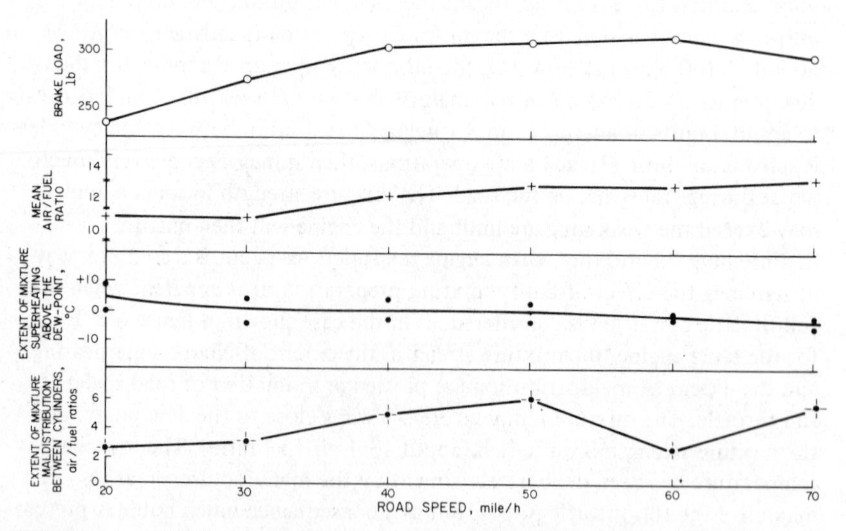

Figure 6.5 Variation of brake load, mean air/fuel ratio, extent of mixture superheating and extent of mixture maldistribution at full throttle (1.7 l engine) (1 mile/h = 1.609 km/h; 1 lb = 0.454 kg)

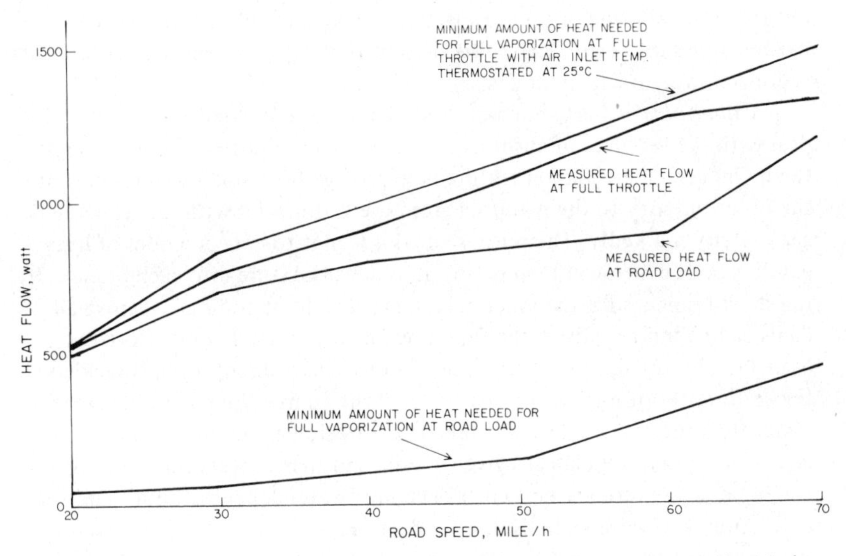

Figure 6.6 Comparison of heat transferred to mixture by the inlet manifold with the minimum amount of heat theoretically needed to vaporize the fuel (1.6 l) (1 mile/h = 1.609 km/h)

6.6 indicates the extent of charge cooling accepted by the manufacturer in order to augment the peak power.

There is no simple answer to the problem of ensuring adequate mixing of fuel and air in the inlet manifold of an engine, and each engine manufacturer adopts a solution that suits him on a cost-effectiveness basis. Many palliatives exist, all of which are usually painstakingly explored with new engines. Only when these have failed will the manufacturer resort to fuel injection or to special quick-heat manifolds[12] to solve specific fuel distribution problems. Some of the palliatives[3,4,13] are worthy of note because they indicate just how difficult it is to prepare a uniform mixture in the inlet system of an engine. These are as follows.

(a) *Carburetter design and orientation* The bias on the liquid stream emerging from the jet or emulsion tube of a carburetter persists along the length of the manifold. The direction of opening of the throttle plate can cause the front two cylinders of an engine to be as much as 30% richer than the rear. Turning the carburetter by 180° can reverse the effect. If this happens, then either the carburetter jetting must be modified or (more usually) a different carburetter design must be tried.

(b) *Inlet silencers* The air intake horn on the inlet silencer is surprisingly critical. It is possible to alter the mixture strength metered by a carburetter and also the mixture distribution by small changes in the air cleaner itself or in the tube connecting the air cleaner to the carburetter. It is usual to carry out the final tailoring of the carburetter to an engine by modifying the balance holes in the air cleaner. At present this aspect of carburetter tailoring is done entirely on an ad hoc basis.

(c) *Manifolds* Certain manifold design criteria[4,13] have been evolved to deal with the problem of ensuring satisfactory distribution of fuel amongst the cylinders. One common feature is to arrange for a small well to exist at the point of entry to the manifold from the carburetter without reducing the gas velocity markedly. The purpose of this well is to collect a pool of heavy gasoline ends which can then be swept as a wet mixture into each branch. If the short branches for the inner cylinders prove to be too rich, then small dams can be built to divert the flow towards the end cylinders. Horizontal branches are usually symmetrical and front-to-back distribution is tackled at the carburetter proper. It is usual to allow the front cylinders to be about 5% richer than the rear to allow for the effects of slow-speed hill climbing. The end turns of manifold always give trouble, and there are several methods for forcing mixture into the end ports. The normal tendency without manifold correction is towards rich inside cylinders and weak outside cylinders. Adequate spirals or dams can be used to correct this. Another method is to use the inertia of the heavy ends to throw the wet fuel out of the mixture stream onto the outside wall of the manifold where it is forced to stay by wall attachment effects until it reaches the port.

It is clear from the foregoing that many engine development problems arise simply from the inability of the designer to handle the air/fuel mixture as a single-phase system. Over the years fairly satisfactory techniques have been worked out by supplying just the right amount of exhaust heat to ensure reasonable mixture distribution amongst the cylinders at road load. Deficiencies in the engine have been compensated for by running at richer mixtures than those absolutely essential. For full-throttle operation, however, the maximum-power condition requires the liquid fuel to provide evaporative cooling, and so of necessity the mixture must be rich.

6.4 Cold Starting and the Use of the Choke

During a cold start all the mixture maldistribution problems normally present in the fully warmed-up engine are exaggerated for the following reasons.

(a) The manifold has to handle relatively large amounts of liquid.

(b) Because of the low engine speed, the gas velocities through the carburetter are low so that mixing is even poorer than usual.

(c) Again because of the low cranking speed, there is little manifold depression to encourage fuel vaporization. A consequence of this is that very rich mixtures must be used for starting and cold running, and these rich mixtures have a serious effect on the engine's fuel economy. Typically, air/fuel ratios[14] of 1/1 or 2/1 are needed to ensure a reliable cold start, and these

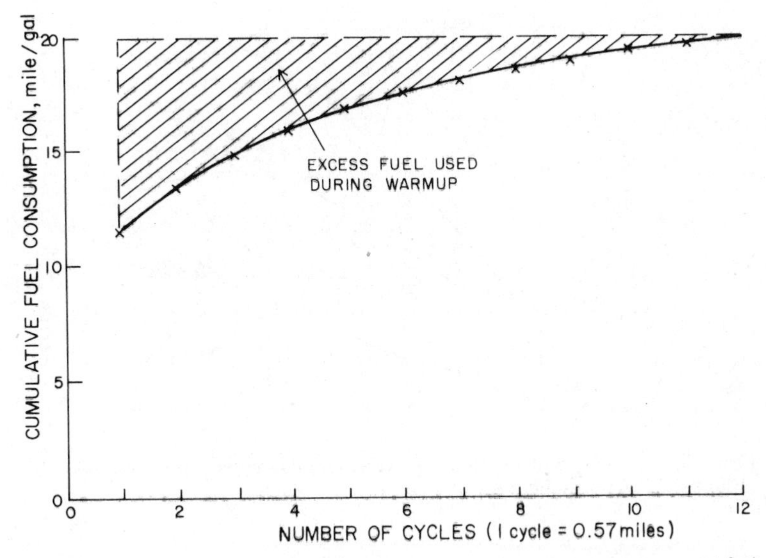

Figure 6.7 Fuel consumption during warm-up at ambient temperature –10 °C[1]
(1 mile/gal = 0.354 km/l; 1 mile = 0.609 km)

are obtained by using the choke. As the engine warms up, so the need for the choke decreases, and it can gradually be released. An example of the effect on fuel economy of the use of the choke during warm-up is given[11] in figure 6.7. During the first cycle when the engine was cold the choke had to be pulled out fully, giving a fuel consumption of about 11 mile/gal (25.7 l/100 km). As the engine gradually warmed up, so the fuel evaporation gradually improved and the choke could be pushed home. Eventually, after 12 cycles or so, the fuel economy improved to about 20 mile/gal. This effect is almost entirely due to the poor quality of the mixture supplied to the engine. Had some form of quick-heating device[12] been in use, then the number of cycles needed to achieve fully warmed-up fuel economy would have been greatly reduced. Ideally and with a perfect mixture the fully warmed-up economy would have been reached after the first cycle.

6.5 Engine Power and its Impact on Fuel Economy During Road Service

A typical gasoline engine will develop 50 bhp/l at full throttle and maximum speed. Thus, a 2 l engine may be capable of developing 100 bhp (74.6 kW) at 5000 rev/min, the corresponding road speed being over 90 mile/h (145 km/h). Such speeds cannot legally be reached on public roads in most countries, and consequently for most of its life the engine runs much more slowly and develops much less power.

Figure 6.8 shows graphically the power output of an engine as a function

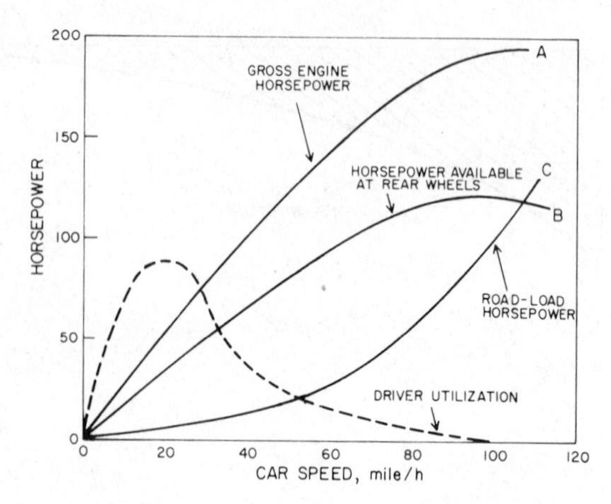

Figure 6.8 Power available to the driver at various speeds (1960 compact car)[15] (1 hp = 0.746 kW; 1 mile/h = 1.609 km/h)

of the road speed of the car in which it is fitted[15]. Curve A shows the gross full-throttle horsepower of the engine and curve B the horsepower available at the wheels after subtracting the transmission losses. At each speed, curve B corresponds to about 65% of A, indicating that transmission and other losses amount to about 35% of the full-throttle engine power. The road load is shown as curve C, and the difference between this road-load curve and the curve B at any road speed is the surplus power available for acceleration. Driver utilization of the available power is also indicated, so that the most important regimes of engine operation for fuel consumption can be identified for the journey studied as being at road speeds between 15 and 30 mile/h (24 and 48 km/h) and at a brake horsepower of about 30 (22.4 kW). This contrasts with the maximum horsepower of 125 bhp (93 kW) theoretically available at the wheels.

Whilst the example quoted is for an American car, the same kind of information on driver utilization can be obtained for any car by using an engine service analyser[16], and figure 6.9 shows typical power and fuel utilization patterns obtained for a 1.8 l car in normal service. The most important operating modes, so far as time spent and also fuel consumed are concerned, are far below the maximum power output of the engine. This is typical of most motorists' driving patterns.

So far as fuel economy is concerned, we are obviously critically dependent on low-load operation, and we need to know just how the specific fuel consumption (sfc) varies with the engine load. This relationship is a complicated one and is usually presented as a map which shows how the brake mean effective pressure (bmep) varies with engine speed for constant values of the brake specific fuel consumption (bsfc). A typical map for a 2.1 l engine is shown in figure 6.10, and the road-load curve for the car in which the engine is fitted is shown on the map as the curve AB. At low road loads the engine is always running at high values of the bsfc, and it is only at high loads that the more efficient engine operating modes are employed.

The problem of efficient operation at low loads is common to all types of prime mover. Figure 6.11 shows[17] how different types of engine compare when the vehicles powered by them are driven over a standard test cycle. The most efficient is the diesel, for which the power control is obtained by varying the mixture strength. In this engine the pumping losses are low

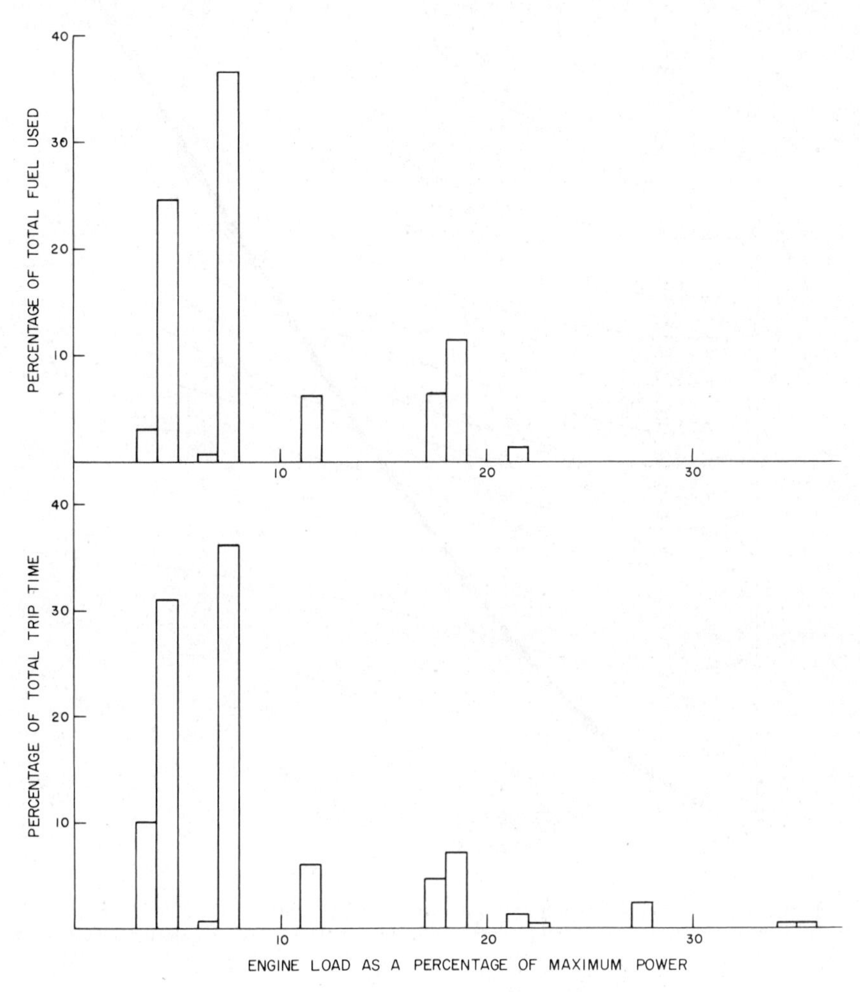

Figure 6.9 Engine and fuel utilization patterns found for typical road service (1.8 l passenger car)

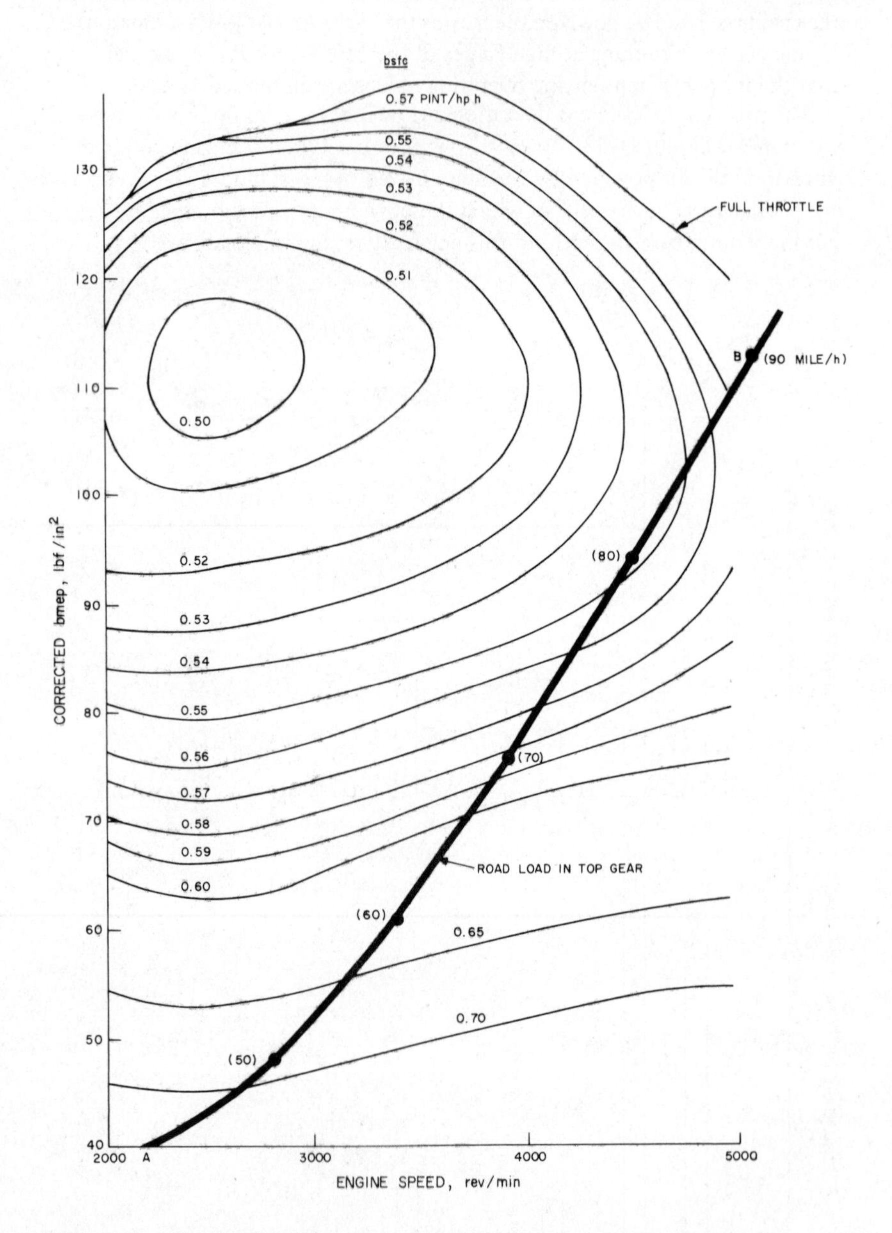

Figure 6.10 Typical fuel consumption map for a 2 l engine with road-load curve superimposed (1 lbf/in^2 = 6.89 kN/m^2; 1 pint/hp h = 0.211 l/MJ; 1 mile/h = 1.609 km/h)

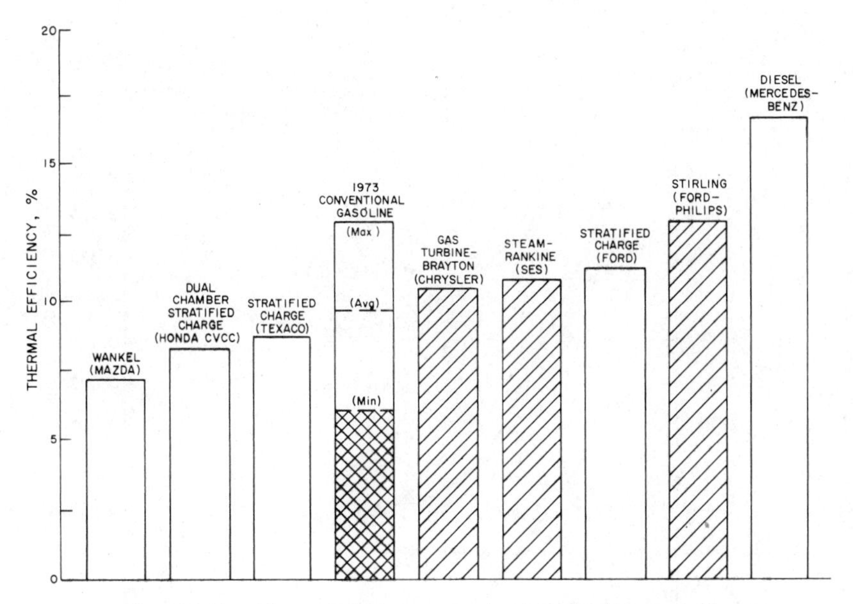

Figure 6.11 Thermal efficiency over the federal driving cycle[17]

because there is no throttle; the compression ratio is high, and the fuel/air mixture is always weaker than stoichiometric. Unless charge stratification[18] is used, gasoline engines cannot presently operate in this way, largely because of mixture preparation problems. It is interesting to note that the MAN L9204 FMV[19] engine (which employs direct injection into the cylinders) will run equally well on both premium gasoline and diesel fuel and will produce the same part-load economy on each.

6.6 Mixture Quality and Fuel Consumption at Part Load

The effect of running engines on well-prepared mixtures can be explored by fully vaporizing the fuel and by supplying it to the engine already mixed with the air at a constant air/fuel ratio. Experiments of this type have been carried out by a number of workers[20-25]. Robison and Brehob[21] carried out a comparative study of the indicated specific fuel consumption (sfc) of a six-cylinder test-bed engine running both with normal carburation and with fully vaporized fuel. Some of their results for the part-load condition are shown in figure 6.12. The most striking observation is the much wider range of air/fuel ratios over which the engine will still run and yet will preserve a low sfc. Steady-state test-bed work, however, only tells part of the story, and in figure 6.13 some comparative results[25] are shown for a car running over a standard test cycle with a normally carburetted mixture and with a constant-strength mixture supplied by a mixture generator. In this case the total fuel

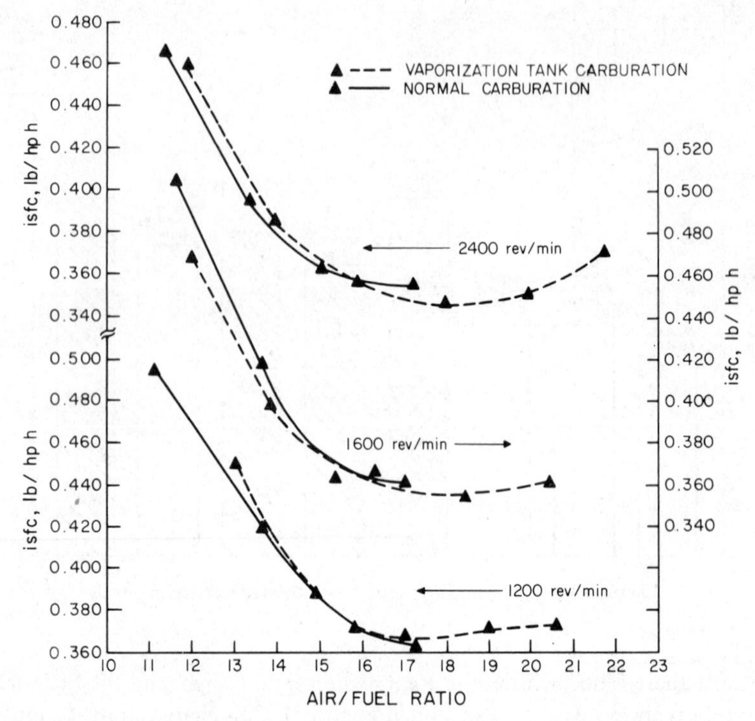

Figure 6.12 Effect of air/fuel ratio on isfc at road load for vaporization tank and normal carburation[21] (1 lb/hp h = 0.169 kg/MJ; 1 rev/min = 0.1047 rad/s)

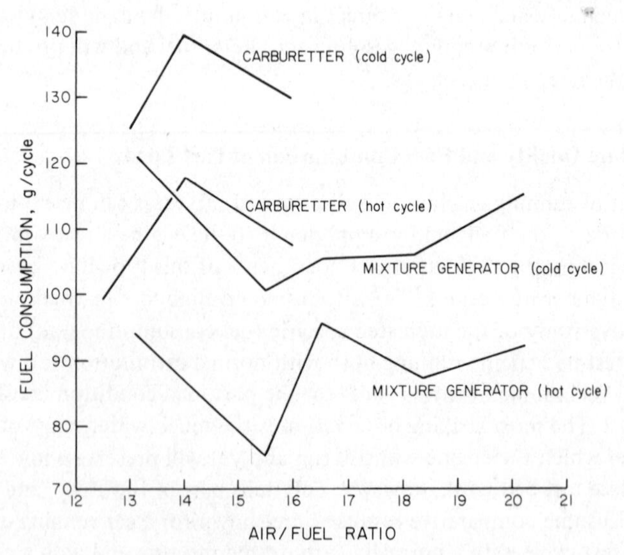

Figure 6.13 Comparison of fuel consumption: carburetter versus homogeneous mixture, cold and hot Californian seven-mode cycles[25]

economy benefits contain a contribution from the better mixture strength control exercised by the mixture generator and, in the case of the cold running tests, from the absence of choked operation during starting.

Clearly substantial fuel economy benefits are possible provided full use is made of the properties of the well-prepared mixture.

6.7 Mixture Quality and Fuel Consumption at Full Throttle

By using a single-cylinder engine to eliminate the effects of mixture maldistribution, Dodd and Wisdom[23] investigated the effect of mixture quality on bsfc at full throttle. Four fuel preparation systems were employed: one normal carburetter, one fuel vaporizer and two different types of fuel injection system. A comparison between the extremes (viz. normal carburetter and fuel vaporizer) is shown in figure 6.14, which takes the form of the characteristic fuel loops for the mixture preparation systems. Complete vaporization of the fuel shifts the whole curve downwards and to the left, indicating three distinct effects.

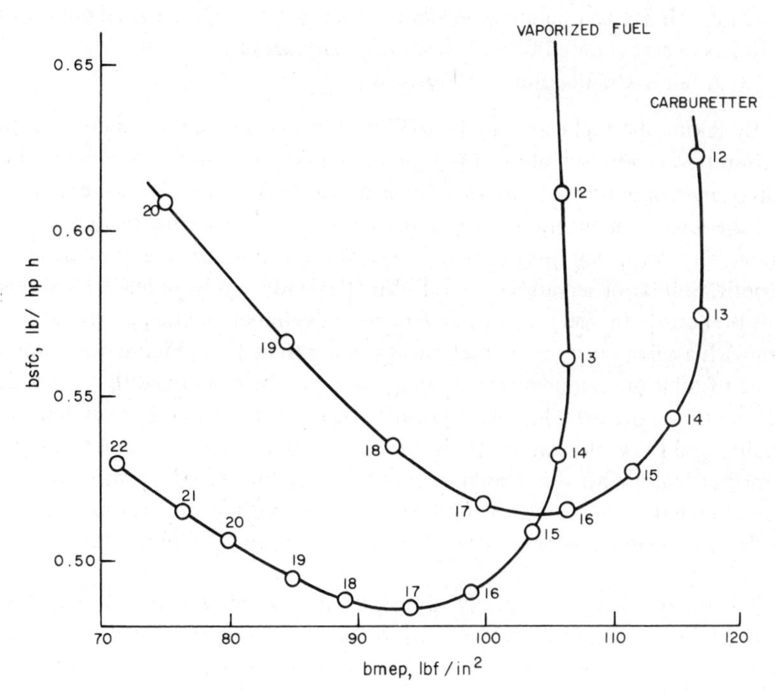

Figure 6.14 Mixture loops comparing carburetted with fully vaporized fuel. Numbers indicate air/fuel ratio (1 lb/hp h = 0.169 kg/MJ; 1 lbf/in^2 = 6.89 kN/m^2)

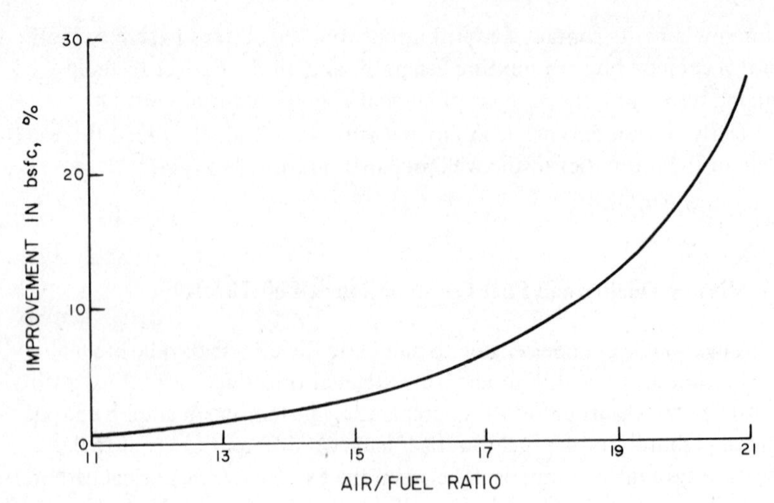

Figure 6.15 Percentage improvement in bsfc with vaporized fuel

(1) A drop in the minimum bsfc, which is more apparent the leaner the mixture becomes.

(2) An ability to operate at leaner mixtures (in the figure the lean-mixture limit has been extended from 20/1 to 22/1 air/fuel ratio).

(3) A fall in the maximum power.

By taking the fuel economy benefits as a percentage improvement over the carburetted engine a result of the type shown in figure 6.15 is obtained. The importance of good fuel vaporization with lean mixtures is clearly revealed, and the improvement increases steadily the leaner the mixture becomes. Conversely, with rich mixtures, improved vaporization provides very little benefit. Similar observations for full-throttle conditions were made by Robison and Brehob[21]. In their case, however, the six-cylinder engine just would not run with a carburetter at air/fuel ratios weaker than 17/1. Figure 6.16 shows some of their measurements. Summing up it may be concluded that (a) with rich mixtures the bsfc does not appear to be critically dependent on mixture quality and (b) with lean mixtures the mixture quality becomes increasingly important, and with very lean mixtures fully vaporized fuel can provide improvements in bsfc of 25% or more compared with normal carburation.

Smaller benefits were obtained with the two types of fuel injection system tested.

Inevitably, in going to a fully vaporized fuel for this example we have lost power by decreasing the charge density. Figure 6.14 shows that the maximum power fell by approximately 9% when going to vaporized fuel. Unless this derating of the engine is accepted we shall not gain the fuel economy benefits

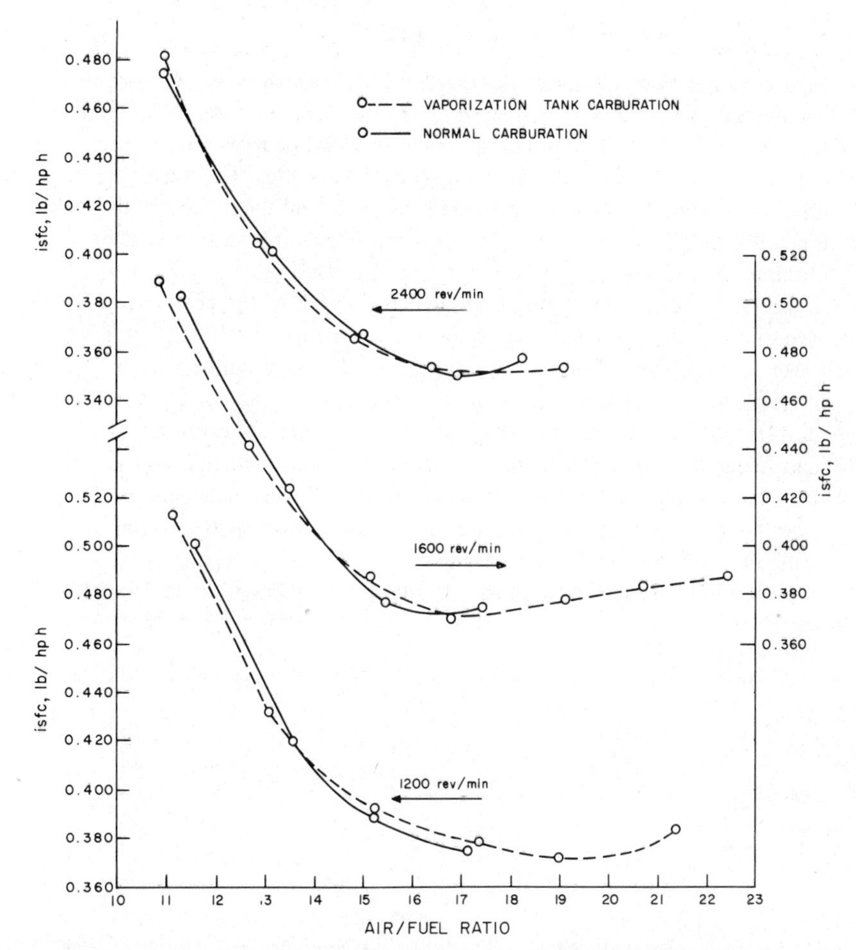

Figure 6.16 Effect of air/fuel ratio on isfc at wide-open throttle (WOT) for vaporization tank and normal carburation[21] (1 lb/hp h = 0.169 kg/MJ; 1 rev/min = 0.1047 rad/s)

that are possible by exploiting the potential of the engine to run on weaker mixtures with a greater thermal efficiency.

It must be emphasized that derating the engine at full throttle by using fully vaporized fuel and weak mixtures is quite different from derating the engine by closing the throttle. By using the fully vaporized lean mixtures we obtain a genuine improvement in the bsfc, and because carburetter design is no longer restricted by the need for fine atomization we can restore some of the power by fitting a lower pressure-drop instrument. If we derate an engine by closing the throttle, there is a worsening of the sfc because of the higher pumping losses and poorer combustion.

6.8 Engine Operation with Weak Mixtures

Figure 6.17 shows how the brake thermal efficiency (bthe) of Otto-cycle engines varies with mixture strength at different compression ratios[26]. The top curve relates to the standard air cycle and lower curves refer to more conventional working cycles involving fuel/air mixtures. The lowest thermal efficiency is obtained with rich fuel/air mixtures and the highest with weak mixtures. Significant thermal efficiency improvements can therefore be obtained by using weak fuel/air mixtures. There is a price to pay for this, however, as indicated in figure 6.18, which shows how the peak engine power varies with mixture strength. Here the weaker the mixture, the poorer is the power output. Since, however, we do not use the maximum power very often on the road it seems desirable to explore the possibilities of using weak mixtures as a means of better thermal efficiency and fuel economy at part load. What stops this? The main problem arises from the need to ensure that the weak-mixture limit is not exceeded in any cylinder of the engine at any time, even during transients, and also to ensure that adequate torque is available from the engine when needed.

It should be noted that figures 6.17 and 6.18 are based on calculated

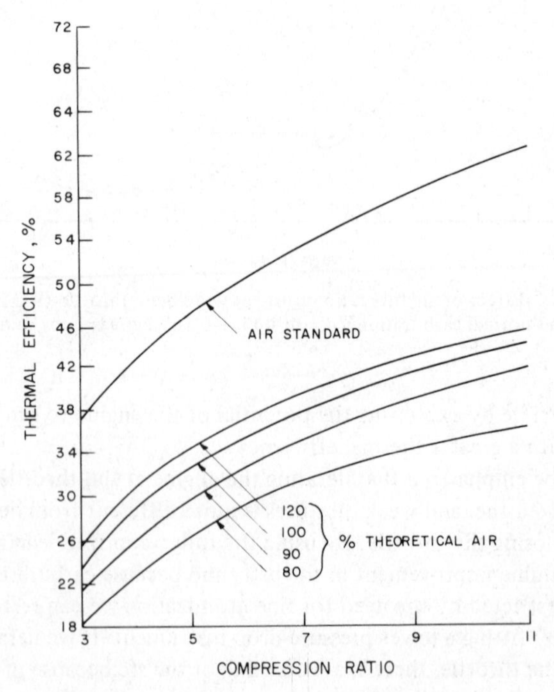

Figure 6.17 Calculated thermal efficiency for the Otto-cycle internal combustion engine[26]

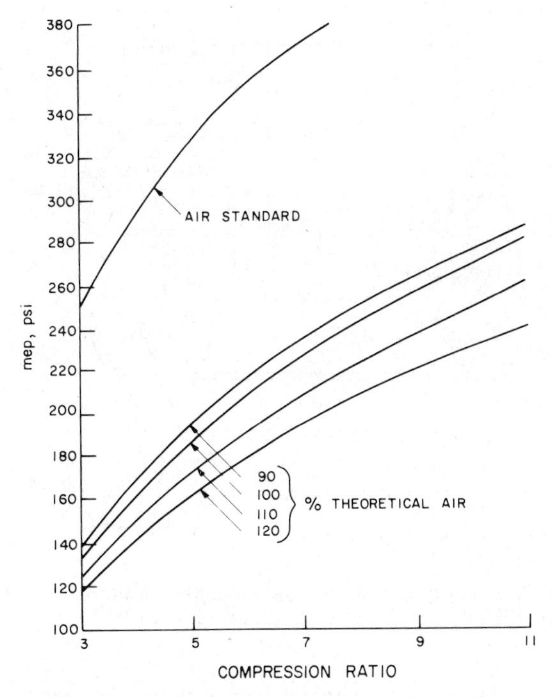

Figure 6.18 Calculated mean effective pressure (mep) for the Otto-cycle internal combustion engine[26] (1 psi/(lb/in^2) = 6.89 kN/m^2)

cycles which treat the combustion process in an idealized way and make no allowance for the effects of spark timing, flame speed, etc. If the spark timing or the spark energy are not optimized then poor results will be obtained in practical engines, particularly when running on weak mixtures. Practical measures can be taken to increase the spark energy so that this is not the factor which determines the weak-mixture limit. An extensive literature is accumulating on this topic[27], and it will not be discussed further here. Optimization of the spark timing is of course vital to the success of any attempt to improve the fuel economy of engines regardless of the mixture preparation system employed.

6.9 Practical Systems for Improved Mixture Preparation

Numerous attempts have been made to develop mixture preparation systems which will allow engines to run reliably on weak mixtures. The subject has been reviewed by Schweitzer[28]. Conventional carburetters and induction systems are not adequate for air/fuel ratios of 17/1 and leaner, so that special systems are required. The best-known systems at present are the Ethyl Corporation 'hot box' manifold[29–32], the Vapipe[11,25] and the Dresserator[33].

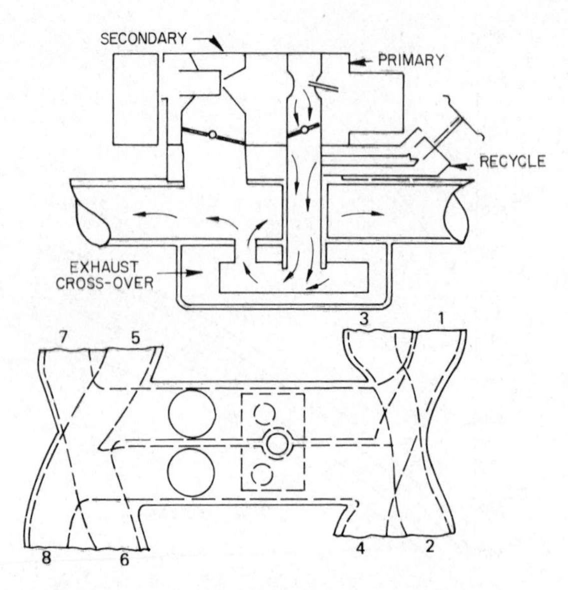

Figure 6.19 Ethyl Corporation's rectangular hot box manifold for a 360 cubic inch displacement (CID) Plymouth[33] (1 cubic inch = 16.39 cm^3)

The Ethyl Corporation 'hot box' system is shown diagrammatically in figure 6.19. A 'hot box' is located under the primary barrel of a Quadrajet carburetter and is sunk into the exhaust manifold crossover of the engine. The fuel/air mixture emerging from the primary barrel passes through the hot box under all driving conditions, but the fuel/air mixtures from the secondary barrels bypass the hot box. High air velocities through the secondary barrels are relied on to give good distribution even without evaporation; the resulting mixture distribution is claimed to be improved for all driving conditions. The main flow of air is not heated with this system and therefore the full-throttle volumetric efficiency of the engine is maintained. The mixture distribution spread claimed for this system is about 3% at idle and 1.6% at 30 mile/h (48.3 km/h), an improvement by a factor of 2 over the original engine.

The Vapipe employs a heat pipe to convey heat from the exhaust system to vaporize all the fuel under all operating conditions of the engine. A general arrangement of the system is shown in figure 6.20. Mixture distribution with the Vapipe is invariably excellent, and table 6.2 summarizes mixture distribution results obtained with a number of Vapipe conversions. These results should be compared with those in table 6.1 which are for the same cars before conversion to Vapipe. Because evaporative charge cooling is absent with Vapipe systems, there is some loss of full-throttle power. It is, however, sometimes possible to regain this power by using a larger carburetter or by improving the engine breathing to exploit the potential of the better mixture distribution.

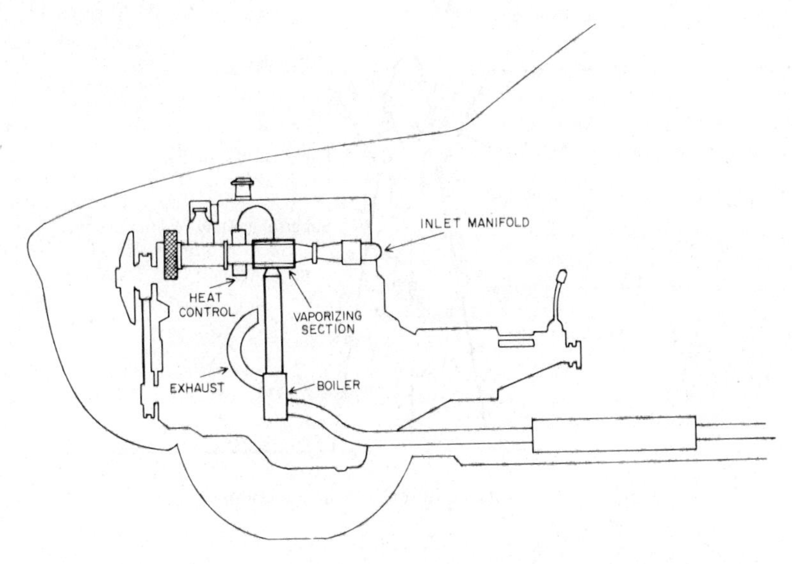

Figure 6.20 Location of Vapipe[25]

The Dresserator is a variable Venturi carburetter with a mechanically actuated fuel distribution bar. The principle of the device is shown in figure 6.21.

Although still in a very early development stage the device is claimed to improve atomization, mixture quality, air/fuel distribution and control. These are obtained as follows.

Table 6.2 Typical spread of air/fuel ratios between cylinders for Vapipe conversions

Engine mode[a]	Maximum air/fuel ratio range between cylinders						
	1.3 l, make A	1.8 l, make A	1.5 l, make B	1.6 l, make B	2.0 l, make C	1.9 l, make D	2.3 l, make E
Idle	0.5	0.8	0.8	0.3	0.6	0.1	0.1
30 mile/h RL	0.1	0.1	0.2	0.1	0.2	0.1	0.3
50 mile/h RL	0.3	0.1	0.5	0.1	0.3	0.1	0.5
70 mile/h RL	0.3	0.3	0.7	0.2	0	0.1	0.2
30 mile/h FT	0.3	0.2	0.5	0.3	0.4	0.3	0
50 mile/h FT	0.3	0.6	0.6	0.3	0.1	0.3	0.3
70 mile/h FT	0.3	0.6	0.7	0.4	0.4	0.4	0.6

[a]RL, road load; FT, full throttle.
1 mile/h = 1.609 km/h.

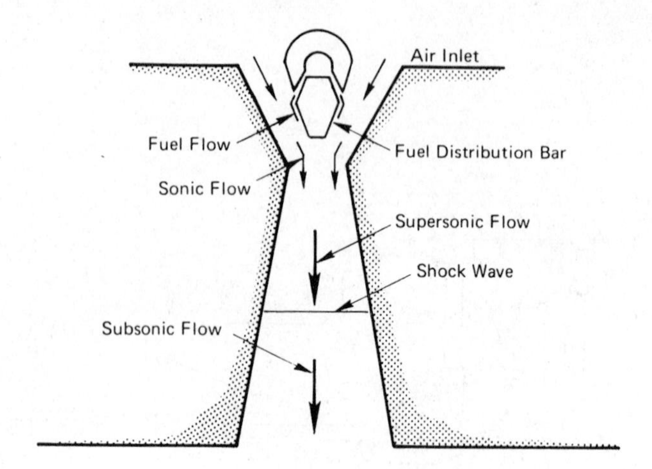

Figure 6.21 Sonic carburetter principle (Dresserator)[33]

(1) By introducing the fuel over a large surface area that is all subjected to sonic flow.

(2) By passing the partially prepared mixture through a shock wave to atomize the fuel and to improve mixing.

(3) By eliminating the flow distortions caused by a downstream throttle plate.

(4) By coupling the throttle with a linear fuel control valve to achieve a constant air/fuel ratio.

The Ford Motor Company[34] have compared the sonic carburetter with other types of mixture preparation system, and their views are indicated in figures 6.22 and 6.23 for various parts of the inlet system. They consider that, after complete fuel vaporization, the sonic carburetter provides the most homogeneous mixtures.

6.10 Fuel Injection Systems

There is much controversy about the benefits of fuel injection systems for gasoline engines. With most of these systems the fuel is injected into the inlet manifold, and only low injection pressures are employed. It is not usual to inject directly into the cylinder, partly because of charge stratification problems and partly because it is more expensive to do so. It is well established that the maximum power output for a given engine can be improved with injection, e.g. with electronic fuel injection (EFI) a 2.8 l Daimler–Benz gasoline engine delivers 185 bhp (138 kW) at full throttle, whereas with a carburetter it will only deliver 150 bhp (112 kW). This advantage is due to the better volumetric efficiency and full-throttle power enrichment with EFI.

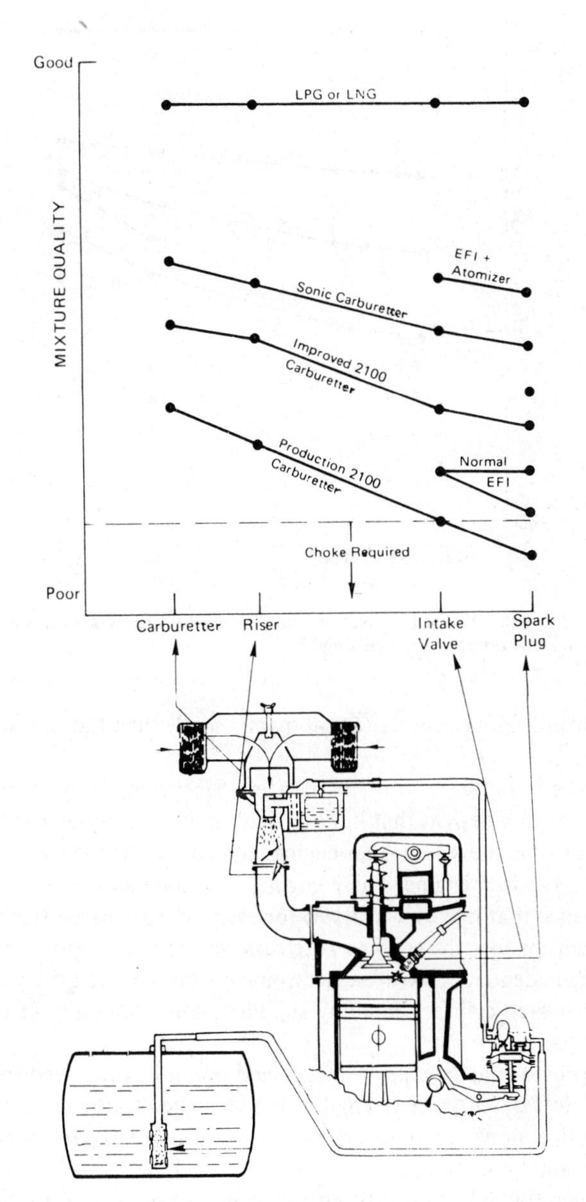

Figure 6.22 Ford Motor Company estimate of induction system mixture quality trends under cold-start and drive conditions[34] (LPG, liquefied petroleum gas; LNG, liquefied natural gas)

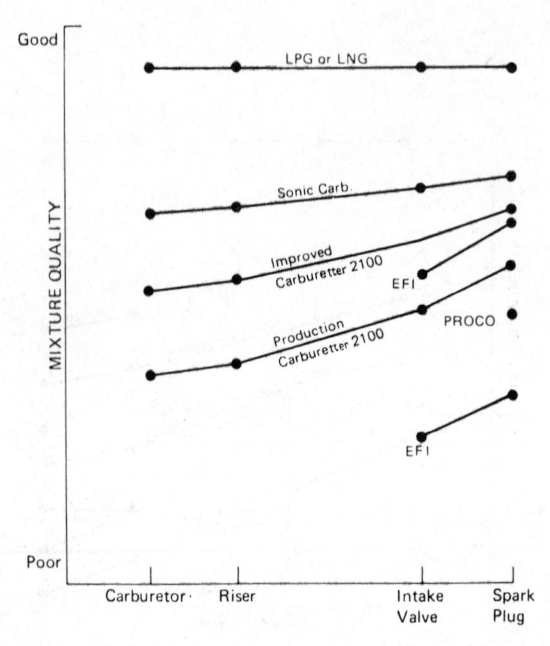

Figure 6.23 Ford Motor Company estimate of induction system mixture quality trends under hot operating conditions[3 4]

(The EFI manifold has a better aerodynamic shape than the carburetter manifold.)

EFI can offer better mixture distribution for poorly designed inlet manifolds or for engine designs that have difficult induction problems. One example of this is the V12 Jaguar engine, and another is the air-cooled four-cylinder Volkswagen engine. These engines have long intake manifolds and a firing sequence that makes it difficult for each cylinder to get the correct charge. Traditionally, the engines were carburetted rich to ensure that all cylinders had adequate mixtures; electronic injection solved the problem by ensuring that each cylinder could by supplied with essentially the correct amount of fuel.

EFI improves cold-start performance and minimizes the need for over-fuelling. Unless the mixture strength can be adjusted to the maximum economy value, however, there appears to be no fuel economy benefit over a properly tailored carburetter.

Even when fuel injection is fitted to an engine the extent of fuel maldistribution that remains can be quite large. The problem has been discussed by Borrmeister et al.[2] A typical specification for a high-pressure fuel injection system permits deviations in the delivery volume by the pump and injectors of 4% at full load, 6% at half load and 10% at idle. Therefore, air/fuel ratios in individual cylinders may vary over these limits simply because of the mechanical tolerances in the pump. A practical investigation by Borrmeister et al. of a typical production pump showed that these tolerances were frequently exceeded. When a low-pressure system is employed, additional difficulties arise from gas evolution in the circulating fuel system. Usually with EFI the quantity of fuel injected is modulated by varying the length of the injection pulse. When this is short, as in the case of the Bosch D Jetronic system (roughly 4 ms at full load), the rise time of the electrical pulse driving the injector (about 1.6 ms) can form a large fraction of the total injection period, and even small differences between the electrical properties of injectors can have a significant effect on the amount of fuel metered. In those injection systems where the injection pulse is not timed to each individual cylinder and where all injectors are fired simultaneously (e.g. the Bosch L Jetronic system), a more accurate control over the injection pulse length is possible, but it does mean that for some cylinders the injection occurs when the inlet valve is closed. It is unlikely that perfect mixture distribution is possible under these circumstances.

Air maldistribution may also occur in some engines[35], and in general it is found that matching a fuel injection system to an engine is every bit as difficult as matching a carburetter.

6.11 Concluding Remarks

(1) The effect of mixture quality on fuel economy is significant and increases in importance the weaker the mixture strength becomes. It is of particular importance for smooth engine operation under transient conditions.

(2) In multi-cylinder engines, good mixture quality eliminates inter-cylinder fuel maldistribution and allows the carburetter to be tuned to the maximum-economy mixture strength.

(3) Under cold running conditions, good mixture quality allows minimal use of the choke, thereby reducing short-trip fuel consumption.

(4) With a perfect mixture preparation system, at least part of the engine power could be controlled on mixture strength alone. This would significantly improve part-load fuel economy.

(5) Many devices exist for improving mixture quality. They improve fuel economy, however, only if the mixture strength and spark timing are adjusted correctly to exploit the benefits that they confer.

References

1. S. P. Skripkin, V. I. Vorobev and P. G. Romanchikov. The influence of the distribution of the vapour and liquid component of the fuel mixture stream on the non-uniformity of mixture components in the engine cylinder. *Automob. Prom.* (March 1974) 12
Also *Motor Ind. Res. Assoc. Transl.*, No. 74-7-190 (1974)

2. J. Borrmeister, F. Drechsler and V. Nghia. Current problems of mixture distribution in four-stroke petrol engines. *Tech. Univ. Dresden, Kraftfahrzeug-Tech*. (December 1974) 364
Also *Motor Ind. Res. Assoc. Transl*., No. 30/75 (1974)
3. G. F. Gibson. Carburation. *Proc. Inst. Automob. Eng., London*, **30** (1935–6) 367
4. A. Taub. Carburation. *Proc. Inst. Automob. Eng., London,* **32** (1937–8) 573
5. J. M. Zucrow. Fuel mixture distribution. *Soc. Automot. Eng. Trans.,* **24** (1929) 162
6. H. T. C. Yu. Fuel distribution studies – a new look at an old problem. *Soc. Automot. Eng. Trans*., **71** (1963) 596
7. D. E. Cooper, R. L. Courtney and C. A. Hall. Radioactive tracers cast new light on fuel distribution. *Soc. Automot. Eng. Trans*., **67** (1959) 619
8. M. H. Collins. A technique to characterize quantitatively the air–fuel mixture in the inlet manifold of a gasoline engine. *Soc. Automot. Eng. Pap*., No. 690515 (1969)
9. L. Eltinge. Determining fuel/air ratio and distribution from exhaust gas composition. *Soc. Automot. Eng. Pap*., No. 680114 (1968)
10. W. D. Mills and G. A. Harrow. A dielectric cell technique for the continuous measurement of fuel/air ratio under transient conditions of engine operation. *Soc. Automot. Eng. Pap*., No. 710162 (1971)
11. R. Lindsay, A. Thomas and J. L. Wilson. Mixture quality, gasoline vaporization and the Vapipe. Paper presented at *Inst. Mech. Eng. Conf. on Power Plants and Future Fuels, January 1975*, No. C35/75
12. W. D. Bond. Quick heat manifolds for reducing cold-engine emission. *Soc. Automot. Eng. Pap*., No. 720935 (1972)
13. F. C. Mock. Design of inlet manifolds for heavy fuels. *Soc. Automot. Eng. Trans*., **15** (1920) 983
14. J. S. Clark. Initiation and some controlling parameters of combustion in the piston engine. *Inst. Mech. Eng., Proc. Automob. Div*., **5** (1960–1) 165
15. G. A. Hoffman. Automobiles today and tomorrow. *Rand Corp. Rep*., No. RM 2922 (November 1962)
16. J. H. Lewis. An engine service analyser for the study of engine usage in road vehicles. To be published
17. J. J. Brogan. Alternative powerplants. *Soc. Automot. Eng. Pap*., No. 730519 (1973).
Also in *Soc. Automot. Eng. Spec. Publ*., No. SP-383 (1973)
18. T. Date, S. Yagi, A. Ishizuya and I. Fuji. Research and development of the Honda CVCC engine. *Soc. Automot. Eng. Pap*., No. 740605 (1974)
19. A. G. Urlaub and F. C. Chmela. High-speed multi-fuel engine L9204 FMV. *Soc. Automot Eng. Pap*., No. 740122 (1974)

20. E. E. Fawkes. The mixture requirements of an internal combustion engine at various speeds and loads. *Thesis*, Massachusetts Institute of Technology (1941)
21. J. A. Robison and W. M. Brehob. The influence of improved mixture quality on engine exhaust emissions and performance. Paper presented at *Western States Combustion Inst. Meet., October 1965*, No. WSCI 65-17
22. J. H. Jones and J. C. Gagliardi. Vehicle exhaust emissions, experiments using a pre-mixed and pre-heated air–fuel charge. *Soc. Automot. Eng. Prepr.*, No. 670485 (1967)
23. A. E. Dodd and J. W. Wisdom. Effect of mixture quality on exhaust emissions from single-cylinder engines. *Inst. Mech. Eng., Proc. Automob. Div.*, **183** (1968–9) 3E
24. R. Lindsay, A. Thomas, J. A. Woodworth and E. G. Zeschmann. An investigation of the influence of homogeneous charge on the exhaust emissions of hydrocarbons, CO and NO from an engine using a continuously generated mixture of gasoline and air. *Soc. Automot. Eng. Pap.*, No. 710588 (1971)
25. R. Lindsay and J. L. Wilson. Heat pipe vaporization of gasoline – Vapipe. Paper presented at *Committee for the Challenges of Modern Society Conf., Ann Arbor, Michigan, October 1973*
26. T. Baumeister and L. S. Marks. *Standard Handbook for Mechanical Engineers*, 7th edn, McGraw-Hill, New York (1967)
27. T. W. Ryan, S. S. Lestz and W. E. Meyer. Extension of the lean misfire limit and reduction of exhaust emissions of a spark ignition engine by modification of the ignition and intake system. *Soc. Automot. Eng. Pap.*, No. 740105 (1974)
28. P. H. Schweitzer. Towards an economical low-pollution automobile. *Fed. Int. Soc. Ing. Tech. Automob. Pap.*, No. B-1-7 (1974)
29. D. A. Hirschler and F. J. Marsee. Meeting future automobile emission standards. Paper presented at *Natl. Pet. Refin. Assoc. Conf., San Antonio, 6–7 April 1970*, No. AM 70-5
30. Earl Batholomew. Potentialities of emission reduction by design of induction system. *Soc. Automot. Eng. Pap.*, No. 660109 (1966)
31. L. Eltinge, F. J. Marsee and A. J. Warren. Potentialities of further emission reduction by engine modification. *Soc. Automot. Eng. Pap.*, No. 680123 (1968)
32. D. R. Liimatta, R. F. Hart, R. W. Deller and W. L. Hull. Effects of mixture distribution on exhaust emissions as indicated by engine data and the hydraulic analogy. *Soc. Automot. Eng. Pap.*, No. 710618 (1971)
33. US Environmental Protection Agency Consultants. *Natl. Res. Counc., Environ. Prot. Agency Rep. on Emissions Control of Engine Syst.*, presented at *Comm. on Motor Vehicle Emissions Commission on Sociotech. Syst., September 1974*

34. Ford Motor Company. Presentation at *US EPA Comm. on Motor Vehicle Emissions, Panel of Consultants on Engine Systems, 16 May 1974*
35. W. R. Brandstetter and M. J. Carr. Measurement of air distribution in a multi-cylinder engine by means of a mass flow probe. *Soc. Automot. Eng. Pap.*, No. 730494 (1973)

7 The Effect of Vehicle Maintenance on Fuel Economy

J. ATKINSON and O. POSTLE

7.1 Introduction

In discussions about fuel economy much emphasis is usually placed on the influence of driving habits, and there is little doubt that attention to such factors can lead to economies. However, the mechanical hardware that constitutes a modern car is and always has been subject to degradation with usage, and so attention needs to be focussed on the fairly substantial savings that can be made by attention to vehicle maintenance.

Some items of importance that are regularly subject to attention in practice are as follows.

(a) Ignition timing (static, vacuumatic, centrifugal).
(b) Contact breaker setting.
(c) Spark plug electrode gap.
(d) Tappet adjustment.
(e) Engine idling speed.
(f) Idling mixture strength.
(g) Brakes.
(h) Fan belt/alternator belt adjustment.
(i) Lubrication.
(j) Thermostat operation.
(k) Air cleaner condition.

Although the above items are generally acknowledged to have an influence on fuel consumption, much of the evidence is of a subjective nature, and to date few factual data have been published. A great deal of the existing information suffers from three shortcomings: firstly, it often applies to the control of emissions rather than directly to fuel economy; secondly, it often applies to US (not European) cars; thirdly, it often is of the more popular type, designed to boost the claims of a particular device or maintenance procedure.

7.2 Thornton Research Centre Tests

To help remedy the above situation, some time ago the Thornton Research Centre (TRC) conducted a series of experiments designed to evaluate some of the more obvious influences of vehicle defects (or maladjustments) upon vehicle fuel economy. For this work a chassis dynamometer was employed, together with an electrically operated driving aid to assist the driver to conform to the prescribed driving requirements. The Economic Commission for Europe (ECE) hot-start driving cycle (ECE 15, see figure 7.1) was used, and fuel consumption measurements were made gravimetrically. Each test comprised 4 cycles of 0.65 mile (1.04 km) per cycle with the vehicle dynamometer fully warmed up (except for the thermostat test).

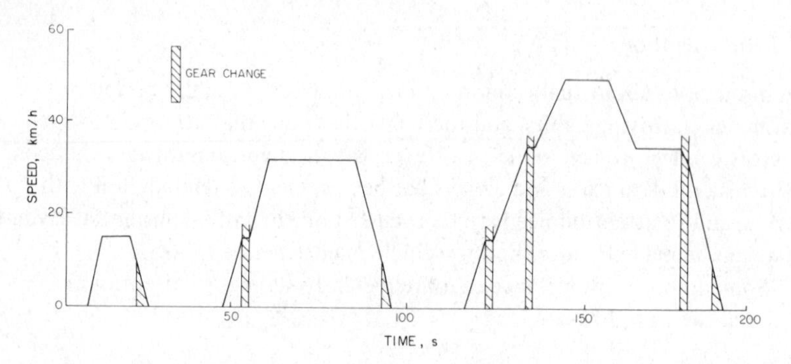

Figure 7.1 ECE 15 exhaust emission test cycle

7.2.1 Deliberate Malfunctions

In these tests six vehicles of model year 1970–3 were deliberately set up to measure the effect on fuel economy of the following.

(a) One shorted-out spark plug (obtained by closing the electrode gap).

(b) Mixture strength, over range of CO emissions at idle from 2% (weak) to approximately 8.5% (rich).

(c) Idling speed, compared between 650 rev/min (low) and 850 rev/min (high).

(d) An inoperative centrifugal advance (obtained by locking the advance mechanism 'bob weights').

(e) Loss of vacuumatic advance (obtained by plugging the vacuum advance pipe from the carburetter).

(f) The thermostat (cold-start tests with and without the thermostat in the cooling system).

(g) A restricted air cleaner (obtained by masking the cross-sectional area of the air cleaner element).

Table 7.1 Fuel economy tests: test cars used and results obtained

Car data		1970 Chrysler Avenger	1971 Vauxhall Viva	1973 Ford Granada	1970 Vauxhall Victor	1970 Rover	1970 Austin
Capacity,	cm^3	1498	1159	2994	1975	1978	1088
Mileage,	miles	40600	50000	6500	50000	49000	45500
Fuel economy, ECE cycle,	mile/gal	24.27	27.09	18.68	19.34	20.74	25.67
	(km/l)	(8.59)	(9.59)	(6.61)	(6.85)	(7.34)	(9.09)
Motor magazine (overall figures),	mile/gal	25.00	29.04	19.00	21.50	22.60	28.20
	(km/l)	(8.85)	(10.28)	(6.73)	(7.61)	(8.00)	(9.98)
Factor		**Change in fuel economy[a], %**					
One failed spark plug		−18.45	−15.02	−4.60	−14.47	−12.29	−10.59
Mixture strength from weak to rich		−9.43		−7.20	−18.92	−6.56	
Increase in engine idling speed from 650 to 850 rev/min		−8.48	−2.87		−5.06	−3.95	
Seized centrifugal advance mechanism			−30.86	+2.40	−17.52	−10.27	−10.82
Failed vacuum advance device			−3.83	−0.16	−7.29	−0.24	−2.26
Removal of thermostat during warm-up		+0.49	−4.94		−1.8		
Restricted air cleaner element			−30.15	−0.53	−0.56	−15.67	−10.59
Binding brake drums			−15.5		−12.87		
Removal of fan blades			+5.24		+3.92	+2.41	

[a] A blank space indicates that no test was carried out. Positive values indicate an increase in economy resulted from the factor examined (i.e. more miles per gallon).

(h) Binding brakes (obtained by running the car with the hand brake applied about two notches and five serrations for the lever and pistol grip types respectively).

(i) Fan blade removal.

Test results on the properly functioning cars, which have been obtained using the ECE 15 cycle (table 7.1) show good agreement with overall fuel consumption values quoted by *Motor* magazine.

Except for the removal of fan blades it can be seen from table 7.1 that most of the maladjustments produced decreases in fuel economy, i.e. lower values of miles per gallon.

The results obtained on the Vauxhall Viva are the most striking, and two comments are necessary.

(1) Of all the cars tested the Viva employed the largest ignition advance, i.e. 28° to 33° before top dead centre (btdc) at 4000 rev/min, which shows the dependence of the engine upon the automatic ignition advance, and the removal of this causes the largest effect of all the cars.

(2) The Viva employed a fixed-jet atmospherically vented carburetter, which is extremely sensitive to a throttled air intake, and this explains the dramatic mixture-richening effect observed when the air cleaner was restricted.

7.2.2 The Sequential Removal of Deliberate Malfunctions – Vauxhall Victor

A series of tests was next made on a Vauxhall Victor in which a number of defects were deliberately introduced simultaneously and were then corrected sequentially. The results are shown in figure 7.2.

From the figure it can be seen that fuel economy in the test cycle fell from 19.34 mile/gal to 11.94 mile/gal when all the faults were present. Subsequently, as the faults were removed, the economy recovered to virtually the original value. The figure also shows the relatively greater importance of vacuumatic advance compared with that of centrifugal advance from an economy point of view and the overriding influence of adjustment of mixture strength. The obvious effect of brake drag is also illustrated; in this car it happened to be fairly severe, but in a real 'in-service' situation it would reflect the degree of maladjustment encountered.

7.2.3 Engine Tuning

Having previously created deliberate malfunctions on vehicles and having measured some effects on fuel economy, attention was then turned specifically to the effects of engine tuning. For this purpose nine vehicles ranging from 1967 to 1973 model years were chosen. All the vehicles had previously

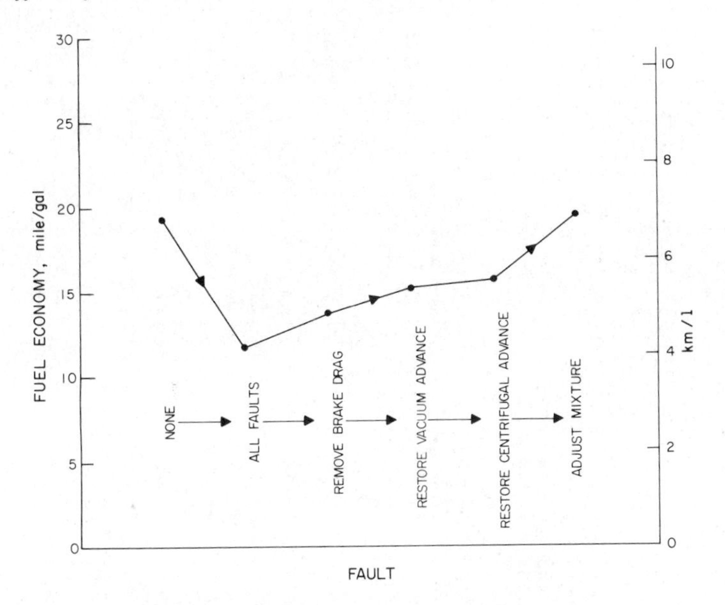

Figure 7.2 Fuel economy test – Vauxhall Victor

been used on research programmes and most had accumulated a fairly high mileage (> 50 000) and had not been recently tuned.

Fuel consumption measurements were made before and after tuning which comprised careful attention to all the following items.

(a) Ignition timing (basic setting).
(b) Dwell angle (contact breaker gap).
(c) Spark plug settings (spark plug gap).
(d) Tappet clearances.
(e) Idling speed and mixture.
(f) Air cleaner cleanliness.

The following separate test conditions were used for comparison.

(i) ECE 15 driving cycle (hot start).
(ii) Makers' recommended idling speed.
(iii) Constant speeds of 50 and 100 km/h road load.
(iv) Full-throttle acceleration from 60 to 100 km/h.

Fuel consumption was accurately measured gravimetrically. The results obtained are shown in table 7.2.

The table gives the results obtained on a percentage basis. There is clearly a tendency to an overall improvement in economy resulting from the tuning procedure adopted. Some negative results, on the other hand, indicate an increased consumption following the tuning exercise.

Table 7.2 Percentage changes in fuel economy following engine tuning

Car	Carburetter type	Change in fuel economy[a], %				
		ECE cycle	At idle	At constant 50 km/h	At constant 100 km/h	During acceleration from 60 to 100 km/h
Ford Cortina 1600	Fixed jet	−0.23	−21.3	+3.77	+5.05	−1.37
Vauxhall Viva 1200	Fixed jet	+3.92	−2.25	+0.77	0	+6.98
Hillman Hunter DL	Variable jet	+5.19	+5.05	−2.33	+16.63	+13.98
Ford Cortina 2000	Fixed jet	+0.96	+2.32	+4.26	+0.14	+6.85
BLMC Austin 1800	Variable jet	+9.65	+19.42	+4.38	+14.52	+18.75
BLMC Austin 1100	Variable jet	+5.88	+12.5	+2.58	+3.06	+5.45
Datsun 160B	Fixed jet	−3.83	−6.97	+1.62	−0.87	+5.44
Mazda RX2	Fixed jet	−0.36	0	0	+1.37	−4.44
Hillman Hunter GL	Variable jet	+2.34	+15.2	−1.38	−3.86	+7.89

[a]Positive values indicate an increase in economy (i.e. more miles per gallon).

Because of the different carburetter types and different tuning techniques, adjustments to idling speed give rise to different overall fuel economy behaviour. For example, the adjustments performed on the cars fitted with fixed-jet carburetters appear to affect only the idling fuel consumption figures – and often adversely. By contrast, cars fitted with a constant-depression type carburetter appear, in the main, to respond to the adjustment of idling speed settings at other parts of the operating spectrum; in table 7.2 the frequent fairly large improvements in fuel economy resulting from the tuning exercise should be noted.

Having thus determined the importance of idling mixture setting on overall fuel economy, a subsequent exercise was conducted on relatively new cars in which the mixture strength at idle was set to a verylow '%-CO-at-idle' setting of less than 4% by means of a Sun 1160 combustion efficiency analyser. These results are given in table 7.3 and show the consistent benefit in terms of economy that resulted from the adjustments made.

From the magnitude of the percentage gains it seems obvious that the greatest improvement in economy was again made in the idling speed range. This appears to be directly related to the technique adopted, i.e. tuning to give minimum CO at idle rather than to give an optimum idling speed setting, as was done in earlier work.

Adjustment to the SU carburetter fitted to the Morris car gave indications of fuel consumption savings throughout the test operating range of the vehicle (very similar to effects noted with the Austin car tested previously) and much greater than those measured on the other makes of car tested.

With reference to the other adjustments carried out, i.e. ignition timing, tappets, air cleaner or plug settings, since these were made collectively, it is not possible to attribute an overriding fuel economy effect to any individual adjustment. Effects resulting from these kinds of adjustments are much more likely to be manifest in startability, knock resistance or 'run-on' behaviour than in changes in fuel economy. The latter may in reality be significant but may tend to go unnoticed by the average driver.

Table 7.3 Percentage improvement in fuel economy resulting from tuning of relatively new cars

Car	ECE cycle	At idle	At constant 50 km/h	At constant 100 km/h	During acceleration from 60 to 100 km/h
1975 Morris Marina 1.8	18.9	44.9	12.42	15.08	6.67
1974 Hillman Hunter	7.2	20.13	1.66	1.6	6.68
1974 Citroën GS 1220	8.4	16.39	7.7	8.9	5.0

TRC work reported above thus shows that the major factors affecting fuel economy are in the following order of decreasing importance.

(1) Idling mixture setting.
(2) Basic ignition timing/dwell angle.
(3) Vacuumatic ignition advance.
(4) Centrifugal ignition advance.
(5) Spark plug condition.

7.2.4 State of Tune of Vehicles in Service

In order to put the above-mentioned factors into perspective in a real situation, we examined the servicing records of 72 vehicles which had been presented for diagnostic performance tests in the Automotive Servicing Laboratory at TRC.

The corrections that had to be made to the vehicles for them to accord with the manufacturers' specifications, and the proportion of the sample requiring attention, are shown in table 7.4.

Table 7.4 Corrections needed on vehicles serviced at the Thornton Research Centre

Correction needed	Percentage of sample
CO at idle	83.4
Basic ignition timing	75
Dwell angle	40.6
Valve adjustment	29.2
Spark plug replacement	23.6
Contact breaker points replacement	20.8
CO at 2000 rev/min	18.1
Cylinder leakage	16.7
Air cleaner replacement	5.6

7.3 Other Similar Test Work

7.3.1 Champion Sparking Plug Company

It is of interest that the Champion Sparking Plug Company have reported[1] on the findings of their 1973–4 series of tune-up diagnostic clinics throughout the UK and Europe (table 7.5).

Although many of the data presented in table 7.4 and 7.5 may at first sight appear to conflict somewhat, it should be noted that, whereas the TRC figures (table 7.4) show the total percentage incidence of a given correction, the Champion data relate to the maximum measurement noted on the items

Table 7.5 European survey of 'tune-up' faults (fourteen countries) carried out by Champion Sparking Plug Company[1] in 1973–4

		Lille	Brussels	Eindhoven	Copenhagen	Malmö	Hamburg	Milan	Barcelona	Athens	Oslo	Helsinki	Vienna	Zurich	UK
Contact breaker gap incorrect,	%	60.2	51.0	38.6	43.8	51.0	35.8	52.8	31.7	50.2	64.0	49.6	50.4	47.5	57.0
Basic ignition timing incorrect,	%	59.2	39.6	52.3	43.8	32.0	15.8	54.7	51.7	58.2	58.3	37.5	50.4	27.7	25.0
Carbon monoxide, $CO > 4.5\%$,	%	54.1	46.2	50.0	56.2	38.9	40.0	68.9	48.0	58.5	52.0	57.4	42.5	38.6	57.2
Spark plugs needing attention,	%	15.0	12.6	27.3	21.9	38.9	8.4	28.3	25.5	3.3	23.4	37.6	10.8	5.9	27.5
Rich mixture at idle,	%	57.6	44.5	37.5	21.9	37.4	42.1	40.6	37.6	52.5	46.9	26.2	30.6	29.7	59.6
Rich mixture at high speed,	%	15.3	44.0	46.6	21.9	17.6	41.0	24.5	26.6	39.5	51.0	48.2	35.6	33.6	57.9
Average number of faults per car (21 test items)		4.7	3.9	3.9	4.0	3.8	2.9	5.0	3.1	4.6	5.7	4.3	4.1	3.1	3.5
Total number of cars									5321						

Table 7.6 Summary of faults diagnosed in over 5000 vehicles in UK towns by Champion Sparking Plug Company in 1975

		Crawley	Gillingham	Southampton	Newport	Coventry	Nottingham	Stoke	Liverpool	Paisley	Stockton
Number of cars		339	516	471	552	546	443	617	608	561	736
Contact breaker gap incorrect,	%	50	61	55	64	57	57	57	66	58	58
Spark plugs needing attention,	%	31	32	36	16	22	28	28	28	26	30
Ignition timing over advanced,	%	29	30	23	23	27	28	22	25	29	22
Ignition timing retarded,	%	9	13	11	16	9	11	11	13	12	14
Carburation rich at cruise,	%	57	60	71	67	56	45	49	49	59	49
Carburation rich at idle,	%	50	62	62	63	61	58	62	52	44	53

specified. In another table published by Champion, relating to cars examined in the UK (table 7.6), a similar pattern of maladjustment in service is depicted. These results show that attention to details connected with both ignition and carburation adjustment was desirable on most of the cars examined.

There is therefore no disagreement about the various factors affecting economy, and those listed in table 7.4 are of major importance from the point of view of achieving maximum fuel economy.

7.3.2 The Importance of Mixture Strength at Idle

Information obtained both at TRC and by Champion Sparking Plug Company shows that it was important to correct the mixture strength at idle on a large proportion of the vehicles that were submitted for tuning. In case it should be thought that idle speed mixture strength has only a minor influence on fuel economy on the road, it is of interest to refer to other work conducted at TRC.

Using a highly instrumented car the fuel consumed at idle was measured under in-service conditions on three different routes (table 7.7).

Table 7.7 Proportions of fuel consumed at idle under in-service conditions

Test route	Percentage of total fuel used at engine idle		
	Max.	Min.	Average[a]
Chester city route 1	35	13	17.5
Chester city route 2	33	11	15.7
Urban commuting	40	4	8.7

[a] Figures represent average values for 46 runs on Chester city routes 1 and 2, and 120 runs on the urban commuting routes.

These results help to put the importance of idling mixture strength into perspective. From the TRC work on the instrumented car, taken as a whole, it may be noted that route characteristics can sometimes have a greater influence upon fuel consumption than does driving habit.

7.3.3 Fuel Economy Benefits from Tuning: US Experience

In the US, Champion Sparking Plug Company have organized a number of practical field tests to identify some effects of engine tuning on fuel economy.

In Riverside, California, using 1964–8 Checker cabs equipped with 230 cubic inch displacement (CID) six-cylinder Chevrolet engines, the average percentage improvement in economy after tuning was found[2] to be 7.5%.

During the winter of 1968–9 at Sparta, New Jersey[2], in a test using residents' cars, tuned cars showed a 5% improvement in fuel economy over untuned cars. The cars in this test averaged 4 years in age and had all been driven for 10 000 miles or 1 year since the last tuning.

More recently Champion have conducted extensive chassis dynamometer tests on 1968 vehicles (emission controlled) to determine the effect of tuning. Substantial reductions were achieved in CO and HC emission, and improvements of as much as 20% were recorded in fuel economy[2].

Shell Oil Company[3] have employed a highly instrumented 1970 model passenger car for measuring road performance and exhaust emissions. In tests under steady-state conditions on a chassis dynamometer the 30 mile/h test speed produced the largest effects in terms of fuel economy.

It was found (1) that a richer idling mixture strength adjustment gave a 5.5% loss in fuel economy, (2) that a plugged positive crankcase ventilation (PCV) valve gave a 3.7% loss in fuel economy and (3) that the composite effect of a 5° spark retard, a plugged PCV valve and rich mixture at idle gave a 12.1% loss in fuel economy.

7.3.4 Tyre Selection Effects

When considering areas of maintenance which might affect fuel economy, there is one which is strictly outside the engine adjustment area but firmly in the replacement area: care in the selection of tyre type for a given vehicle. The work of Crum and McNall[4], also in the US, shows how a 10–12% improvement in economy may be gained by using radial-ply tyres instead of bias-belted tyres. Also, some chassis dynamometer work at TRC some years ago clearly demonstrated the beneficial effects of using radial-type tyres instead of cross-ply on the brake horsepower and acceleration performance of Ford Corsair cars. Table 7.8 lists the findings. Although fuel consumption measurements were not specifically made in this test, the gain in power and reduction in time spent at full throttle whilst accelerating from 30 to 60 mile/h indicate the extent to which work is lost in tyre friction. Similarly, the Michelin company in the UK[5] have noted that an 8.5–9% improvement in fuel consumption was obtained on Cortina and Marina cars simply by changing from cross-ply to radial-ply tyres.

Concerning the maintenance of vehicle tyre pressures, it is well known that underinflation can be very deleterious to fuel economy and tyre wear. Thus it has been shown[4,6] that a 25% decrease in tyre pressure gives rise to a significant decrease in fuel economy (of the order of 5–10%, depending on driving conditions) and a reduction of 25% in tyre life.

Table 7.8 Effect of radial-ply and cross-ply tyres on brake horsepower and acceleration time

Car	Tyre type fitted	Maximum bhp at 60 mile/h full throttle	Time to accelerate from 30 to 60 mile/h at full throttle[a], s
Ford Corsair V4 (Car 1)	Cross-ply	40	26.5
	Cross-ply	42	24.6
	Radial-ply	45	22.8
Ford Corsair V4 (Car 2)	Radial-ply	41	26.2
	Cross-ply	42	24.6
	Radial-ply	43	23.6
Specification requirement		40	27.9

[a]Average of three runs.

7.4 Concluding Remarks

(1) Work done by Shell Research Limited and others shows that engine-tuning factors have a major influence on fuel economy and that correct mixture strength settings and conformity with motor manufacturers' ignition specifications are particularly important. The major maintenance factors affecting fuel economy appear in the following order of decreasing importance.

(a) Idling mixture strength and engine idling speed.
(b) Basic ignition timing/dwell angle.
(c) Vacuumatic ignition advance.
(d) Centrifugal ignition advance.
(e) Spark plug condition.

(2) The full benefits with respect to fuel economy can only be achieved by systematic and precise diagnosis of engine malfunctions followed by corrective adjustments. For this purpose modern diagnostic facilties are required, and both CO and HC meters are essential, used in conjunction with accurate instrumentation for setting ignition timing.

(3) Each vehicle exhibits its own unpredictable pattern of engine faults and malfunctions. The incidence of the various faults appears in the following general order of decreasing frequency.

(a) Idle mixture setting.
(b) Basic ignition timing.
(c) Dwell angle.
(d) Valve adjustment (tappets).
(e) Spark plug replacement.

The correction of each of these faults does not necessarily give the same fuel economy benefit.

(4) Comprehensive information on the frequency of occurrence of each maintenance item is slowly being built up, but similar information on the severity of the faults is lacking, and it is therefore not yet possible to make a general statement about the amount of each correction that should be expected to be made in service.

(5) The magnitude of fuel consumption will, of course, in each particular case depend on the mechanical state of the vehicle. Some vehicles with many defects may show fuel economy benefits as large as 20% after the proper corrections have been made to provide satisfactory functional performance. However, the published work suggests that an improvement of up to 6% would be more typical. The exact magnitude, spread and durability of this improvement require further study.

(6) With evolving car design, new equipment, particularly for emission control, is being added to cars. Faults that can occur (e.g. valves, exhaust gas recirculation (EGR) systems, heated air intake, etc.) then pose new and as yet largely unexplored problems of vehicle maintenance with respect to fuel economy.

References

1. Champion Sparking Plug Company. *Motor Mail International*, Barrett Card Associates Ltd, London (1974–5), p. 11
2. Champion Sparking Plug Company. Tuned You Win – Untuned You Lose, *Service Corner* (September–October 1973)
3. W. I. Doty and L. J. Olejnik. A mobile instrumented vehicle for measuring road performance and exhaust emission. *Soc. Automot. Eng. Pap.*, No. 720212 (1972)
4. W. B. Crum and R. C. McNall. Effects of tire rolling resistance on vehicle fuel consumption. *Tire Sci. Technol.*, **3**, No. 1 (February 1975) 3.
   Also *Am. Soc. Test. Mater. Comm. F9 on Tires, Symp. on Tires and Fuel Econ., May 1974*
5. *Motor* (12 July 1975) 61
6. C. E. Scheffler and G. W. Niepoth. *Soc. Automot. Eng. Pap.*, No. 650861 (1965)

8 The Effect of Emission Controls on Fuel Economy

D. R. BLACKMORE

8.1 Introduction

The last seven years have seen a progressive lowering of permitted vehicle emission levels, particularly in the US (table 8.1) where signs of a levelling-off are perhaps beginning to be evident. While this was going on, scientists within the industry were noting what sacrifices in performance were needed to achieve these levels, and one of these was fuel economy. When the fuel crisis of 1973 developed, the question of the deleterious effect of emission control on fuel economy was raised more sharply than before, and a public debate ensued in the US in particular.

On the one side was the motor industry who wished to point out its difficulties in being forced to meet stringent new emission standards with minimal lead time, and how this situation had led inevitably to a fuel economy penalty (e.g. a 22% loss in fuel economy between 1967 and 1973 in a typical Ford car, 60% of which was attributed to emission controls[1]).

On the other side were the motor industry's critics, for instance the US Environmental Protection Agency (EPA) spokesmen[2,3] who stated that the degradation in fuel economy between pre-1968 cars and 1973 cars attributable to emission controls was only 10.1% in sales-weighted terms and that small cars (3500 lb (1590 kg) inertia weight or less) had actually shown an improvement. Although their overall sales-weighted figure was not very far removed from Ford's estimate, they went on to state[2] that 'there is no inherent correlation between emission levels and fuel economy' or again[3] 'no law of thermodynamics links engine efficiency and exhaust emissions'.

The debate had also been joined sometime earlier by the oil companies and the lead-additive companies in connection with the voluntary move by the motor industry to lower compression ratios to accommodate the use of unleaded fuel of research octane number (RON) 91, a change that was said by Ford[1] to decrease fuel economy by 3–5%, by Du Pont[4] by 8% but by EPA[5] by less than 3%.

As with most such public debates, the main beneficial effect has been the stimulation of the various parties into producing some very relevant and

Table 8.1 Exhaust emission legislation in US for light duty motor vehicles (as of June 1976)

	Federal (49 state) standards			California standards		
	HC, g/mile	CO, g/mile	NO_x, g/mile	HC, g/mile	CO, g/mile	NO_x, g/mile
1967 (uncontrolled)	15	90	4–6		The same as federal	
1968	6.3	52	–	6.3	52	–
1969	6.3	52	–	6.3	52	–
1970	4.1	34	–	4.1	34	–
1971	4.1	34	–	4.1	34	4.0
1972	3.0	28	–	2.8	28	3.2
1973	3.0	28	3.1	2.8	28	3.0
1974	3.0	28	3.1	2.8	28	2.0
1975	1.5	15	3.1	0.9	9	2.0
1976	1.5	15	3.1	0.9	9	2.0
1977 (statutory)	1.5	15	2.0	0.41 (proposed)	9 (proposed)	1.5 (proposed)
1977–9 (proposed EPA)	1.5	15	2.0		The same as federal	
1978 (statutory)	0.41	3.4	0.4[a]		The same as federal	
1980–1 (proposed EPA)	0.9	9.0	2.0		The same as federal	
1982 (proposed EPA)	0.4	3.4	0.4[a]		The same as federal	

All numbers quoted in terms of 1975 CVS-CH procedure (the CVS-CH procedure is a cold-start constant-volume sampling procedure with the 1972 US urban driving dynamometer schedule; the first 505 s of the cycle is repeated).

Values for the years 1972, 1973 and 1974 were originally set in terms of 1972 CVC-C procedure (the CVC-C procedure is a cold-start constant-volume sampling procedure with the 1972 US urban driving dynamometer schedule). Values for the years 1970 and 1971 were originally set in terms of the California seven-mode cycle, and those for 1969 and earlier were quoted in concentration terms for this cycle.

Conversions have been made to a common basis, but there must be some uncertainty with them, especially for the NO_x data.

[a]The 0.4 g/mile NO_x standard is being suggested as a 'research objective' in recent proposals.

interesting data which we shall attempt to survey here. The reported work has fallen into two categories: (a) historical surveys of the performance of vehicles (either that of the whole US new car fleet in a most valuable EPA programme or that of individual or small groups of typical representative cars) in attempts to pick out the trends and influencing factors; (b) attempts to attribute fuel economy gains or losses to individual exhaust emission control measures. The latter procedure inevitably becomes very complex because of the many interactions between the different control measures, which makes their cumulative effect far from additive.

Anticipating the conclusion, we shall show that the crux of the debate turns not only on whether a particular fuel economy measure and/or emission control device can be made to work in the limited development time available (though this is difficult enough and the US motor industry has had a remarkably successful record) but also on whether the customer is prepared to meet the cost in both financial and performance terms. To put it in other words, will increasingly improved technology to some extent break the correlation between fuel economy and emission control; and will the customer buy it, and when? These questions are currently being faced by US legislators as they consider the wisdom of setting fuel economy legislative levels in addition to exhaust emission ones, what magnitude they should be, and when they should become operative.

8.2 Historical Surveys of Vehicle Fuel Economy

8.2.1 US Experience

The main data source that is currently being built up is the EPA annual fuel economy survey of all new cars, the data for which come from essentially the same chassis dynamometer test as do the exhaust emission certification data. There have to date been three annual publications for model year 1974[5], 1975[2] and 1976[3] cars, and EPA surveillance data[5] have been used to cover earlier years back to 1957.

The size and weight of the vehicle were quickly and hardly surprisingly recognized as important factors controlling fuel economy. Figure 8.1 shows how fuel economy varied with inertia weight for the 1976 model year. Two features are notable: (1) the spread in fuel economy at a given inertia weight is very large, indicating that other factors also contribute strongly; (2) the use of alternative technology (e.g. the diesel or the gas turbine) can give data outside the limits of gasoline-powered cars.

If the data for the whole fleet are pooled, then the trend of the average fuel economy can be plotted against the year. Figure 8.2 shows how the gradual decline between 1967 and 1974 was sharply reversed for 1975 and 1976 model years. This whole trend is relatively unaffected if it is plotted on a sales-weighted basis or on a fixed model mix basis, indicating that any

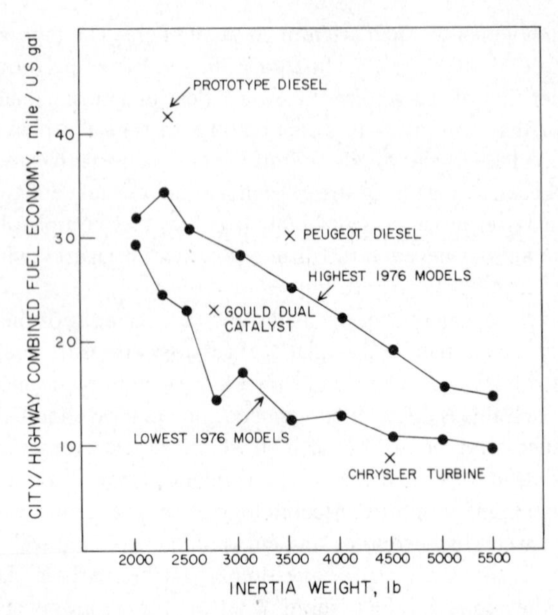

Figure 8.1 Fuel economy versus inertia weight for 1976 gasoline-powered cars compared with other vehicles[3] (1 mile/US gal = 0.425 km/l; 1 lb = 0.454 kg)

trends in the sale of smaller new vehicles has not yet shown itself. It is worth noting that the improvement in sales-weighted fuel economy since the 1974 model year now stands at 27%, and, since this 1974 datum was taken as the basis for the US President's voluntary 40% fuel economy gain by 1980, the target is already two-thirds achieved. Moreover General Motors, who some say had a poor fuel economy performance in the 1974 model year, have already achieved a 39% gain by 1976, and this is in large measure due to the advent of exhaust catalyst technology that permits recalibration of their engines for greatly improved economy.

Further analysis[3] of the EPA data on an inertia-weight basis results in the very interesting data (figure 8.3) for the improvement relative to 1957–67 vehicles, i.e. before the introduction of any exhaust emission legislation. It is evident that the improvement in economy for small cars has been maintained (but has not been increased), whereas the deterioration for large cars has been converted into a gain. Austin et al.[3] offer the plausible explanation that smaller cars, by virtue of their smaller exhaust volume, have been able to meet the legislative limits for mass emission using less stringent and less fuel-consuming control methods. Since the European market is mainly made up of such smaller cars, this result is of great interest and probable relevance there. The control measures that have been mainly employed in these cars have been a leaning of the air/fuel mixture to meet CO and HC limits, coupled with only a limited amount of spark retard or exhaust gas recirculation (EGR) to meet

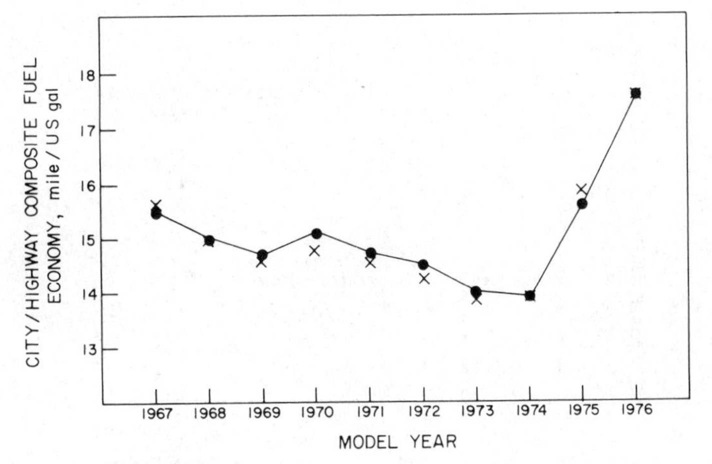

Figure 8.2 Sales-weighted fuel economy trends for 1967–76 (x indicates fixed model mix basis)[3] (1 mile/US gal = 0.425 km/l)

the as yet relatively mild NO_x limits. As long as European emission legislation is also relatively mild, one can also expect to see a small gain in fuel economy, and particularly so since most European manufacturers are learning how to develop their technology in the US arena first.

A more detailed statistical analysis of the 1976 EPA data has been undertaken by Murrell[6] with a view to understanding the factors that influence the data. It emerged that the most practically significant factors for large cars are cubic inch displacement (CID), inertia weight (IW) and drive-train ratio N/V in that order and that for smaller cars N/V is the most significant. These data are shown plotted in figure 8.4; however, it should be borne in mind that the effects are not additive, since correlations exist in the car population

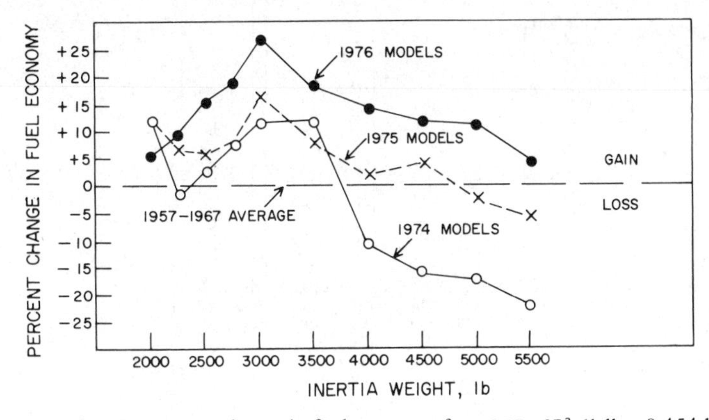

Figure 8.3 Percentage change in fuel economy for 1957–67[3] (1 lb = 0.454 kg)

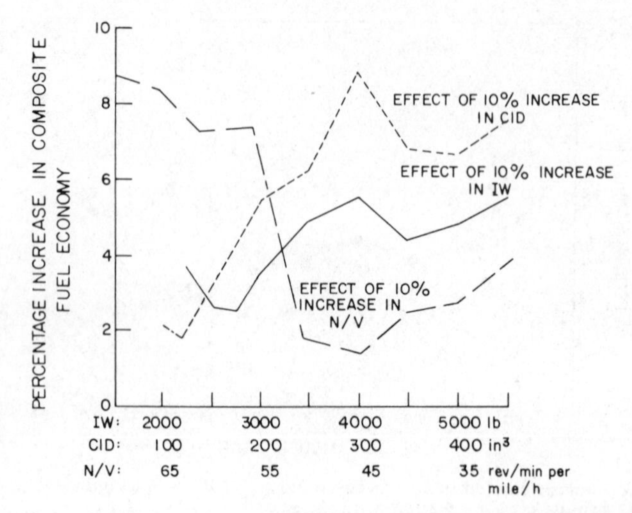

Figure 8.4 Percentage effects of changes in CID, IW and N/V on composite fuel economy for 1976 model year US cars[6] (1 lb = 0.454 kg; 1 in^3 = 0.0164 l)

between these variables (and particularly strongly for CID and IW). Murrell concludes that CID x N/V is the most significant determinant of fuel economy for which a theoretical basis does exist.

This same paper[6] also goes on to summarize the data for relative city fuel economy over the last 3 years, distinguishing between the US federal (49 state) and Californian populations. Figure 8.5 shows how the fuel economy has improved, and, because the same model cars were selected,

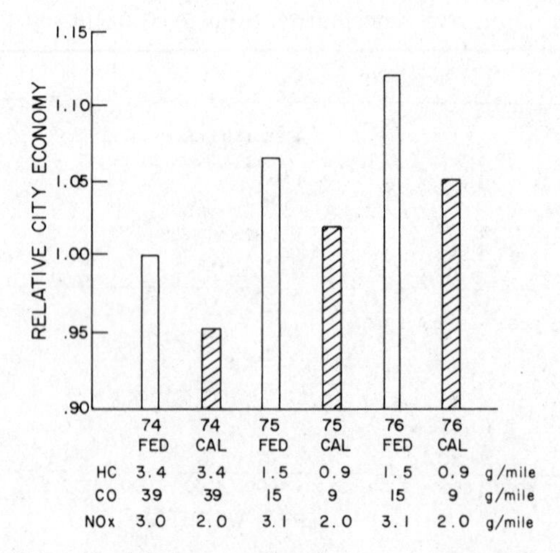

Figure 8.5 Normalized fuel economies of six groups of certification vehicles[6]

these gains are attributable to technological improvements rather than to model mix effects. The conclusions were (1) 'that, with technology fixed at a point in time, fuel economy penalties can be associated with tighter emission levels' and (2) that fuel economy can improve as advances in technology are incorporated. We can see from figure 8.5 that the rate of technological improvement has indeed been most impressively rapid, and much of this has been attributed to the advent of catalyst technology, which even for the 1975 Californian vehicles was still being used by less than half the vehicle manufacturers. Whether this same high rate of improvement in technology can be continued for many years in the future against the evolving background of tightening (and new) exhaust emission legislation is the all-important question. Certainly, the improvement from the use of catalysts will not be repeated in the future, because the major effect was not due to the catalyst per se but was due to the spark timing, which was able to be advanced back to nearly its optimum position.

There have been a number of tests carried out by people other than EPA in order to find out trends in vehicle performance, especially but not entirely with respect to fuel economy. Du Pont[7] have recently reported on a test with six full-sized US cars. The cars were chosen so that a continuity of model could be compared with 1970 – only one of the six cars suffered an engine displacement change. The cars were tested for performance (acceleration time from 0 to 60 mile/h) and for fuel economy on several different cycles (EPA urban and highway, Du Pont city–suburban). When corrected for weight and axle ratio differences the acceleration times were seen to increase by 20% for 1975 cars as compared with their 1970 baseline. Fuel economy suffered an 8–10% loss over the same time span, which Du Pont note is close to that predicted for the 1.5 compression ratio loss in this time. They also note that, if performance had been maintained, the fuel economy loss would have been considerably worse. Du Pont comment that their 1975 data do show a 2–5% improvement in fuel economy over the 1974 data, but this increase is much smaller than the one reported by EPA of 11.5% attributable only to systems changes. Their data are also somewhat in conflict in another respect, namely that EPA[3] report a 7% fuel economy gain from 1970 to 1975, and it seems doubtful that weight and axle ratio corrections would be enough to convert this into an 8–10% loss as found by Du Pont. It must be admitted that the EPA data base is much more extensive, and so these discrepancies point to the difficulty of reconciling conclusions from experiments which involve necessarily choosing a limited number of cars of a given model with statistical analyses of less well co-ordinated information.

In an earlier report, Du Pont[4] attempt to bring together four separate studies, one by themselves and one each by Ford[8], Chrysler[9] and EPA[10], to apportion out the 25% increase in fuel consumption found for representative full-sized cars between 1968 and 1973. They conclude that 7% of this 25% is attributable to weight increases, 3% to increased engine size, 8% to reduced compression ratio and 9% to other emission control measures. A recent paper[1] by LaPointe of Ford gives reasonable agreement with this apportioning. From a study of a fleet of eight car lines from 1967 to 1973, and an analysis of the sales-weighted data for all years (reduced to the 1967 sales weighting and thereby removing the effect of sales differences), LaPointe shows that a 22% loss in fuel economy has been suffered and that 13–14% is attributable to measures taken to reduce exhaust emissions, the remainder being attributable to vehicle weight and engine displacement increases. Interestingly, LaPointe notes a small increase in fuel economy for 1969, a year when relatively mild CO and HC standards were operational and when carburetter adjustments were made in order to give leaner mixtures. He also notes that in 1973 the imposition of NO_x control led to the largest single fuel economy loss of any year under study. LaPointe also classes losses due to decreased compression ratio as related to emission control. He notes that the drop was from 9.3% in 1967 to 8.3% in 1973 and that this theoretically would cause a fuel economy penalty[11] (see also chapter 3) of 6–8% at constant performance or 4–5% if performance were allowed to deteriorate (as it probably was).

This raises another important though partially hidden aspect of this whole story, for such small-fleet studies as Ford's or large-scale annual new-car surveys as EPA's have not dwelt on car performance. Du Pont's work, as mentioned above, does monitor one aspect, namely full-throttle accelerations, and this has noticeably worsened by 20–24% for 1975 compared with 1970 and even 9–11% compared with 1974. However, this is not the only aspect of performance nor even the most important from the customer's viewpoint. Driveability is a measure of the way the car responds to the driver's wishes (see chapter 4), and, particularly over the cold-start and warm-up parts of vehicle operation, this driveability has deteriorated over the last few years. This is attributed both to the need to release the choke quicker than before to avoid excessive cold-start CO emission and also to the advent of EGR systems for NO_x control. The industry has experimented with exhaust heated 'early fuel evaporation' (EFE) systems and with highly volatile fuel. As EGR systems are improved and as carburation and spark advance are better tailored to the needs of the engine, so the driveability will probably improve somewhat. Nevertheless, this limitation on performance is a real one and is an important aspect of the trade-off price for getting better emissions and fuel economy.

8.2.2 European Experience

It is greatly to be regretted that neither annual studies of the EPA type nor fleet studies such as Du Pont's have been carried out, as far as is known, in Europe, although plans are fairly well advanced towards converting the European (Economic Commission for Europe ECE 15) emission test (hot-start version) to such a purpose. Therefore, very little is known about trends

Table 8.2 Summarized exhaust emission legislation outside North America for new gasoline-engined motor vehicles (as of June 1976)

Country	Data	HC, g/km	CO, g/km	NO_x, g/km	Test method
ECE standards	1973	3.2[a]	45[a]	–	ECE 15
	1975 (Oct.)	2.7[a]	36[a]	–	ECE 15
	1976 (Oct.)	2.7[a]	36[a]	3.0–4.5[b]	ECE 15
Australia Sweden		Plan to depart from ECE standards in 1976 and to adopt 1973 US federal methods and standards			
Germany	1976	Proposes further reduction in ECE CO and HC limits			
	1980	Proposes 90% reduction in HC, CO and NO_x over 1970 levels			ECE 15
Switzerland	1975 (Oct.)	2.7[a]	36[a]	2.5–4[a]	ECE 15
	1978 (Jan.)	1.6[a]	22[a]	2.0[a]	ECE 15
	1982 (Jan.)	0.65[a]	9.0[a]	0.60[a]	ECE 15
Japan	1971	–	2.5%v	–	Japanese four-mode
	1973	2.94	18.4	2.18	Japanese CVS
	1975 (April)	0.25	18.4	1.6	New Japanese cycle
	1976 (April)	0.25	2.1	0.84–1.2	Ten-mode cycle

[a]ECE standards are set as a function of car weight: these are for cars of weight 1250–1470 kg.
[b]For cars of weight 910–1810 kg.

and factors influencing fuel economy for cars tuned for the European market and subject to European emission legislation. Table 8.2 shows this legislation and how it has only very slowly tightened up in most European countries. (Japan's legislation is also shown, and, as can be seen, it is planned to tighten up very considerably. The pattern of events in Japan is therefore likely to be similar to that already described in the US, but, even so, few data on the trends in fuel economy in Japan are yet available.)

One very interesting recent comparison by Daimler–Benz is available[12] for a Mercedes V8 engine built initially for the 1975 European market and later modified for the 1975 US market. This 4.5 l engine in its European version has an output as defined by the Society of Automotive Engineers (SAE) of 215 net hp (160 kW) at 4750 rev/min, while the US version has 180 SAE net hp (134 kW). There is therefore a 16% power loss attributed to alteration of the combustion chamber quench zone (resulting in slower

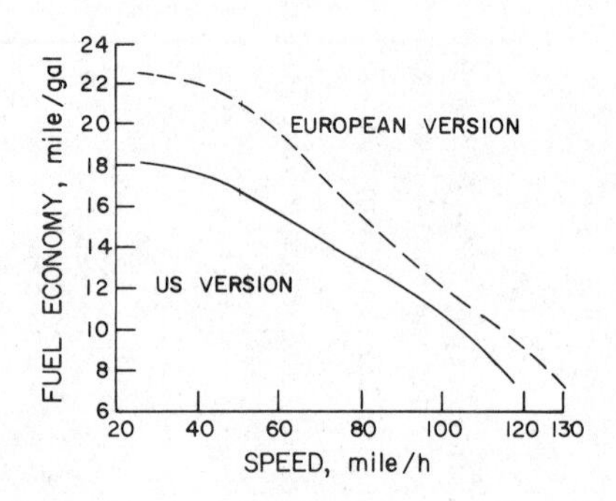

Figure 8.6 Constant-speed road-load fuel economy of 4.5 l engine[12] (1 mile/US gal = 0.425 km/l)

combustion), to lower valve overlap and increased exhaust back pressure due to the presence of the catalyst (resulting in lost volumetric efficiency) and to a reduction in compression ratio from 8.8 to 8.0 to permit the use of 91 RON unleaded fuel. Other modifications for emission control purposes were the use of proportional EGR, improved ignition timing by a double-acting diaphragm on the distributor, and prevention of air injection to the catalyst when the engine is cold. Figure 8.6 shows how the constant-speed road-load fuel economy is degraded in the US version at all speeds but more markedly at the low speeds. Over cyclic driving conditions the difference ranged from 8 to 13%. Measurement of engine efficiency at 4000 rev/min agreed well with this, falling from 29% for the European version to 26% for the US version, a drop of 10% which was identified as arising from increased heat loss to the exhaust. This fuel economy decreased 'despite all efforts to the contrary'.

8.3 The Effect on Fuel Economy of Individual Exhaust Emission Control Measures

It is necessary to realize that the control of emissions, when considered in detail, involves the control of at present four separate pollutants and that measures to control one pollutant will not necessarily control others and indeed may cause increases while at the same time having a variety of effects on other engine variables such as fuel economy, performance, etc. In order to help untangle the extremely complex web that develops in this subject, we shall first of all list the different pollutants, with brief comments on the means used for their control at varying levels of severity, and then we shall list the very many control methods that have been used in recent years, summarizing their cost-effectiveness and consequences on fuel economy and other performance variables.

8.3.1 Exhaust Pollutants

8.3.1.1 Lead

There has been protracted discussion in Europe over the need to control exhaust emissions of lead, and, in spite of the paucity of medical evidence that the material as emitted from the exhaust of a motor vehicle presents a real hazard to human health, the various European governments are proceeding with the European Economic Community (EEC) to enact limits on these emissions by controlling the quantity of lead in gasoline. Recently, the EEC have been considering a directive to set a 0.40 g/l maximum on both premium and regular grades of gasoline by January 1976 and another 0.15 g/l maximum for regular grades alone by 1 January 1978, although the dates, if not the whole question, are still subject to debate. The consequences of such a change for the fuel economy of cars will not be very marked in terms of miles per gallon, but there will be significant effects at the refinery in both money cost (capital as well as running costs) and energy costs. These matters are dealt with more extensively elsewhere (see chapter 3): suffice it to say that these emission benefits do not come without a crude oil penalty, which in refinery terms amounts to some 4% in the short term and 1.3% in the longer term[13], though this latter figure will have to be bought by the investment of as much as $440 (1975) million into refining in the EEC for a total gasoline output of 80 million tons per year. If a lead reduction to 0.15 g/l is called for, these cost figures will increase by a factor of about 3.

In the US a rather different philosophy has been adopted. Unleaded fuel has been legally required in one grade only (91 RON) so that catalyst cars, available in production for the 1975 model year for the first time, may be able to operate with satisfactory durability. They thereby avoid the rather difficult medical question concerning the hazards of lead and put their emphasis on the control of CO and HC emissions. Gradually, as new catalyst cars come on the market, the unleaded gasoline will grow in proportion, and so total lead in gasoline will decrease, aided by further legislation on lead contents in other grades. The energy cost of all this is two-fold. At the refinery, to make 91 RON/83 motor octane number (MON) unleaded pool gasoline will cost less than 1% in energy loss and about 2 cents more per gallon to the consumer[14]. For cars to accomodate the low octane number, the 8:1 compression ratio vehicles will suffer about a 6% fuel economy penalty compared with their predecessors at 9.5:1. This example serves to illustrate again how acutely the octane quality of gasoline in reality affects overall fuel consumption.

An alternative method for reducing lead emissions that has attracted a good deal of interest is the lead trap. The consequences of this device on fuel economy are very minor, but of course there would be no significant penalty to be met at the refinery.

8.3.1.2 Carbon Monoxide

The emission of this pollutant is strongly dependent on the air/fuel ratio that

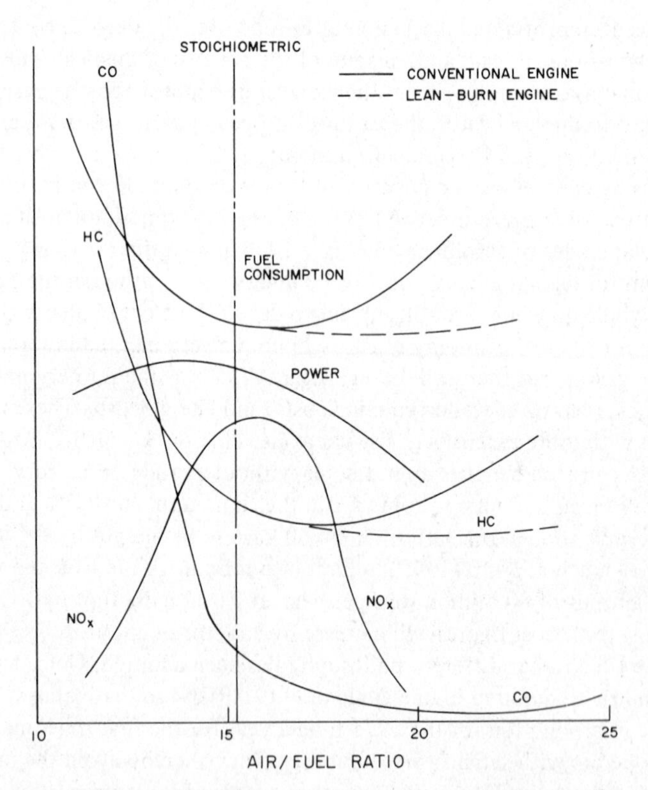

Figure 8.7 Relative relationships of typical engine emissions and performances to air/fuel ratio

the carburetter meters to the engine (see figure 8.7). Measures to control it involve the move to leaner air/fule ratios, and so engine design features that permit this move are much under study currently, especially as they also result in a fuel economy gain. More severe control requires after-burning of some sort, either thermally or catalytically with air injection into a reactor downstream of the exhaust port, but the richness of the mixture in the engine and the extra power needed to operate the air pump both contribute to a fuel economy loss.

8.3.1.3 Hydrocarbons

HC emissions are dependent on air/fuel ratio (but only up to a point, beyond which they increase again, see figure 8.7) and on the degree of spark retard[15] (figure 8.8), a measure which is easy to incorporate but unfortunately very

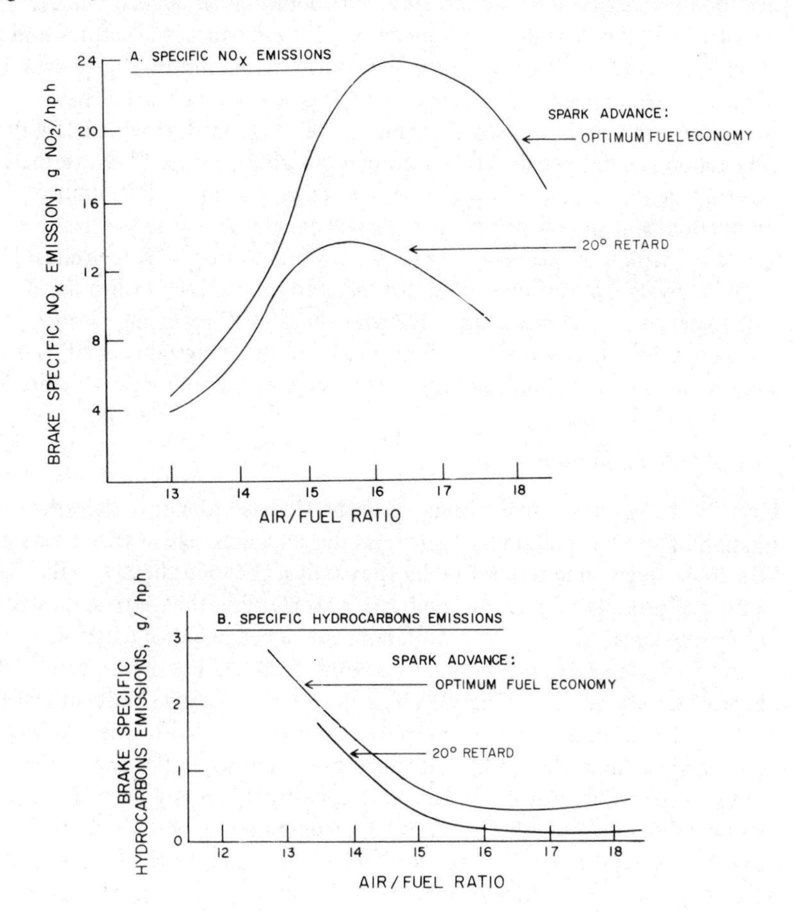

Figure 8.8 Variation of specific NO_x emissions and specific HC emissions with air/fuel ratio and spark advance[15] (1 g/hp h = 0.372 g/MJ)

damaging to fuel economy. More severe control requires after-burning similar to that for CO mentioned above.

8.3.1.4 Nitrogen Oxides

NO_x emissions are rather more complex and difficult to cope with. They increase at first with increasing air/fuel ratio but decrease at still higher values (figure 8.7). They are also very sensitive to spark retard, being greatly reduced similarly to HC emissions[15] (figure 8.8) and strongly dependent on throttle position. Apart from spark retard, a common measure used for their control is EGR, 10% of which can roughly halve the NO_x emission. However, the use of simple EGR gave rise to penalties in fuel economy and also in HC emissions, largely because the EGR metering system was rather crude and gave too much exhaust gas under light load conditions. It appears that, in order to cover the driveability problems and HC emissions, the carburation was enriched and the spark was retarded. However, an improved proportional exhaust gas recirculation (PEGR) system has now been developed that allows the carburation and spark settings to be unchanged, in which case the fuel economy is not penalized, in accordance with prediction[16]. Nevertheless it is still true that severe EGR does cause a restriction on the lean limit of combustion, and so very low NO_x levels still cannot be reached with this procedure without a fuel economy penalty. More severe control requires the use of catalysts, but unfortunately they need to reduce chemically the NO_x to nitrogen and so are not compatible with the CO/HC oxidizing catalysts unless considerable complexity is built in (dual-catalyst systems or 'three-way' catalysts which depend on carefully controlled stoichiometric gas mixtures).

8.3.1.5 Sulphur as Sulphates

Recently the use of oxidizing catalysts in the US has brought to light the emission of another pollutant of concern, the sulphates. These arise because the sulphur in gasoline is oxidized by the catalysts beyond the SO_2 which is the normal product of gasoline combustion to SO_3 and thence to sulphuric acid or sulphates. At present, information on the behaviour of catalytic systems with respect to this pollutant is being gathered, but measurement of it is proving very difficult. The EPA have promised legislative limits in 1979 but have already indicated that this pollutant poses a threat to the continued use of catalysts unless it can be controlled by suitable modification of the catalyst systems. Because of the acknowledged fuel economy benefit that the use of catalysts permits, this legislative control indirectly poses a severe threat to the fuel economy of the motor car. If the alternative route were to be taken for control of sulphur in gasoline at the refinery by means of the energy-expensive hydrodesulphurization process, an alternative and maybe even greater threat to crude oil economy would be posed. For instance, the cost to reduce sulphur in US gasoline to 100 ppm would be $(2–4) x 10^9 (1975) and would require 4–6 years' lead time. This would entail a 1% energy cost and 1–2 cents/US gal gasoline cost increase. As yet this is a wholly US problem, though Japan too may face it within the next few years.

8.3.2 Exhaust Emission Control Systems

To survey this field in any detail is too long a job for the present discussion, especially when to do justice to each system one needs to assess not only its effectiveness in controlling emissions and its consequences for fuel economy but also its consequences for all other relevant aspects, such as driveability (particularly for lean mixture and EGR systems) or durability (particularly for catalyst systems). However, table 8.3 has been drawn up in an attempt to summarize the situation in semi-quantitative terms. The appropriate values for gains in emissions, economy or cost are reached by a combination of the author's judgement coupled with a most valuable survey by the Committee on Motor Vehicle Emissions of the US National Academy of Sciences (NAS)[14].

This table surveys the technological scene as it stands at present. The main message is that in general 'you pay for what you get', and, if this is to be both very low emissions and much improved economy, then 'you pay more still'.

In constructing table 8.3 it proved rather difficult to assess the more powerful and more expensive systems lower down the table, mainly because it is found practically necessary to include some or all of the minor systems listed above but also because such systems can be tuned for several different optima. The Ford Programmed Combustion (PROCO) system is a good example of this, including as it does electronic fuel injection (EFI), oxidation catalysts (OCATs) and EGR, whereas the Texaco system goes still further and utilizes turbocharging as well as exhibiting a multi-fuel capability (in itself a fuel saver in refinery terms).

Likewise the 'lean-burn system' that has been promoted with considerable promise by Chrysler combines the use of lean mixtures and better mixture preparation with electronic control (by a new mini-computer or micro-processor) of the spark advance (ESA)[17]. As inputs to this ESA, there are ambient air and coolant temperatures, throttle position and interestingly also a rate of change of throttle position. They and others hope to develop these ideas of electronic control further to include choke, fuel metering, EGR and transmission and to provide yet further control inputs that would include temperatures and pressures in and outside the engine.

Exhaust emission control system costs to the consumer are summarized by the NAS in table 8.4; it should be noted that for European legislation only five of the first six items seem likely to gain acceptance in the near future, and the combined cost of these amounts to $50 (excluding the evaporative loss system) for a four-cylinder car. This cost would seem well worthwhile to manufacturer and customer alike, especially when a significant fuel economy

Table 8.3 Effectiveness of various emission control systems

	Control system	Comment	Reduction in emission of			Fuel economy gain	Approximate cost increase of hardware change
			CO	HC	NO_x		
(1)	Mixture preparation aids						
	Improved carburetters	Variable-jet multi-venturi, ambient condition controls	S	S	S	S	S
	Intake air temperature control		S	S	0	S	VS
	Sonic carburetters	Dresserator, Ford	S	S	M	S	S
	Petrol injection – electronic		S	S	S	S	M
	– mechanical		S	S	S	S	S
	$+O_2$ sensor feedback	Stoichiometric control (mainly for use with three-way catalysts)	S	S	S	0	L
	Vaporizing – partial	EFE, crossover heat control valves,	S	S	0	S	S
	– total	Vapipe	M	M	M	M	M
(2)	Exhaust gas recirculation						
	Simple EGR		0	(M)	M	(M)	S
	Proportional EGR		0	(S)	M	0	S
(3)	Combustion modification						
	Spark timing retard	Vacuum retard, transmission-controlled spark system	(S)	M	M	(M)	S

Improved ignition system	High-energy ignition, more durable and useful with lean-burn systems	S	S	S	S	S
Compression ratio increase		0	(S)	(S)	M	S
Combustion chamber design		S	S	S	S	S
(4) Exhaust treatment						
Manifold burning + air pump	For HC, CO	S	S	0	(S)	S
Thermal reactor + air pump	For HC, CO	M	M	0	(M)	M
Oxidation catalyst + air pump	For HC, CO	L	L	0	M	L
Reduction catalyst	For NO_x	0	0	L	(S)	L
Dual catalysts	For HC, CO, NO_x	L	L	L	0	VL
Three-way catalyst	For HC, CO, NO_x	L	L	L	0	VVL
(5) Alternative internal combustion engine design						
Stratified charge – torch ignition	e.g. Honda (CVCC) – economy tune	L	M	M	S	L
	– low NO_x tune	L	M	L	(M)	VL
– open chamber (+ oxidation catalyst + EGR)	e.g. Texaco (TCCS) and Ford (PROCO) economy tune	L	L	M	L	VL
	low NO_x tune	L	L	L	0	VL
Diesel		L	L	M	L	L
Rotary	Normal	0	(M)	M	(M)	S
	+ thermal reactor	M	M	M	(M)	M
	+ lean carburation and EGR	M	M	M	0	L

S, small effect; M, medium effect; L, large effect; V, very; the bracketed letter indicates effect in opposite direction.

Table 8.4 Typical component retail prices used in developing vehicle costs for US cars[14]

Component	Retail price[a], $ (1974)		
	Four cylinder	Six cylinder	Eight cylinder
Air injection pump, valves, piping	27	31	33
Fuel evaporation control	8	10	11
Distributor assembly	7	14	21
PCV valve	2	2	3
Improved EGR	11	12	14
Most advanced EGR	18	20	22
OCAT pellet converter	–	–	76
OCAT monolith converter	47	58	–
Reducing catalyst pellet converter	–	–	82
Reducing catalyst monolith converter	63	74	–
Oxygen sensor	4	4	4
Three-way catalyst pellet converter	80	92	92
Fuel injection nozzles	25	42	57
Electronic emissions control unit	50	56	57

[a]Investment allocation is not included.

Table 8.5 Comparative emission control cost data for various systems and emission levels based on intermediate six-cylinder US vehicles[14]

Emission level[a] and vehicle system	Mile/US gal	Increase in lifetime cost, $ (1974)				Discounted life total cost[b], $ (1974)
		Retail price	Fuel	Maintenance	Total	
3.9/33/6						
Base	13.2	0	0	0	0	0
3.0/28/3.1						
Modified	12.1	51	296	325	672	557
1.5/15/3.1						
Modified	12.4	78	210	325	613	512
Lean-burn system	13.9	110	−164	200	146	134
OCAT	13.5	123	76	100	298	265
0.9/9.0/2.0						
Modified	12.0	87	325	325	737	617
Lean-burn system	13.6	110	−96	200	214	190
OCAT	13.1	167	178	113	458	405
Dual catalyst	13.5	249	75	75	399	371

Table 8.5 (Continued)

Emission level[a] and vehicle system	Mile/ US gal	Increase in lifetime cost, $ (1974)				Discounted life total cost[b], $ (1974)
		Retail price	Fuel	Mainte-nance	Total	
Three-way catalyst	13.9	326	−11	12	327	326
CVCC	13.3	210	−25	325	510	449
CCS	15.4	230	−661	12	−419	−310
Diesel	16.5	149	−713	−75	−639	−503
0.4/3.4/2.0						
Lean-burn system	13.3	120	−25	200	295	259
OCAT	13.1	193	177	113	483	431
Duel catalyst	13.1	249	177	75	501	455
Three-way catalyst	13.6	326	51	12	389	378
CVCC	13.1	210	24	325	559	490
CCS	15.4	230	−661	12	−419	−310
Diesel	16.5	149	−713	−75	−639	−503
0.9/9.0/1.0						
Modified	10.8	87	724	.325	1136	948
OCAT	11.8	167	555	112	834	718
Duel catalyst	13.3	249	125	75	449	413
Three-way catalyst	13.7	326	27	12	365	356
CVCC	13.2	209	−1	200	408	352
CCS	14.8	273	−556	37	−246	−158
Diesel	15.1	167	−477	−25	−335	−250
0.4/3.4/1.0						
Duel catalyst	12.9	249	230	75	554	500
Three-way catalyst	13.5	326	75	12	413	398
CVCC	12.2	209	267	200	676	575
CCS	14.5	273	−499	38	−188	−111
Diesel	15.0	167	−458	−25	−316	−234
0.4/3.4/0.4						
Duel catalyst	12.3	331	401	75	807	724
Three-way catalyst	13.2	377	151	38	566	531
CVCC	10.6	215	799	225	1239	1057
CCS	13.3	273	−251	38	60	94

[a]Emission of $HC/CO/NO_x$ in g/mile.
[b]There is considerable uncertainty in these estimates which is about ± $100 for the 1.5/15/3.1 target rising to ± $350 for the 0.4/3.4/0.4 target (for conventional engines with catalysts).

gain is likely to accompany such changes. If ways can be found, through careful design of engine and carburetter, to control CO and HC without an air pump, this incremental cost would be halved and the fuel economy improved all the more.

When it comes to the longer-term future, the NAS report provides a most interesting analysis, given in table 8.5. The table shows the different design options available for achieving different emission control levels and analyses the fuel economy likely to result from its implementation along with its lifetime incremental cost. It is most interesting to note the commanding lead that the diesel enjoys in both fuel economy and lifetime cost*, until the prospect of 0.4 g/mile NO_x is reached. At this point the diesel drops out of the picture and the CCS (controlled combustion system) using a direct fuel-injected stratified-charge engine, e.g. Ford PROCO, Texaco Controlled Combustion System) (TCCS) takes over. For the European prospect of milder emission controls (say 0.9/9.0/2.0 g/mile for HC/CO/NO_x) with lead-containing gasoline, it is interesting to see that the lean-burn systems come out favourably, considerably ahead of even the compound vortex controlled combustion (CVCC) systems, chiefly because of their better fuel economy and lower maintenance costs. Of course, such systems should do somewhat better than indicated in table 8.5 because of their use of higher compression ratios with the higher-octane leaded fuels that will continue to be available in Europe. (However, it should be remarked that other factors may come in to limit compression ratio and octane requirement: in the case of the Honda CVCC at present, the limitation appears to be rather unusually the HC emissions.)

8.4 Discussion

If we return to the question 'by how much is the fuel economy of a vehicle related to the level of exhaust emissions imposed', we see that the motor manufacturer is working within constraints that are certainly more than just two dimensional. A recent paper by General Motors[15] brings together such factors as vehicle weight and engine displacement, in addition to allowable exhaust emission level. Figure 8.9 shows the fuel consumption–vehicle performance curves for vehicles of varying weight and for two emission levels. (Such curves are universally found with engines as currently designed and are

*These values for the diesel should be treated with caution, since they are not necessarily based on consideration of engines of identical performance and since the price of diesel fuel was taken as 42 ¢/US gal compared with 43 for leaded gasoline, 42 for unleaded and 40 ¢/US gal for controlled combustion system (CCS) fuel. These (tax-free) price differentials may not hold for other countries nor even for other refining situations. A recent UK report[18], which also comes out in favour of the diesel engine for European emission levels, puts the retail price increase for an engine of equivalent performance higher than the NAS US estimate (i.e. at around £100), while for an engine of equal capacity the price increase is half this (£50).

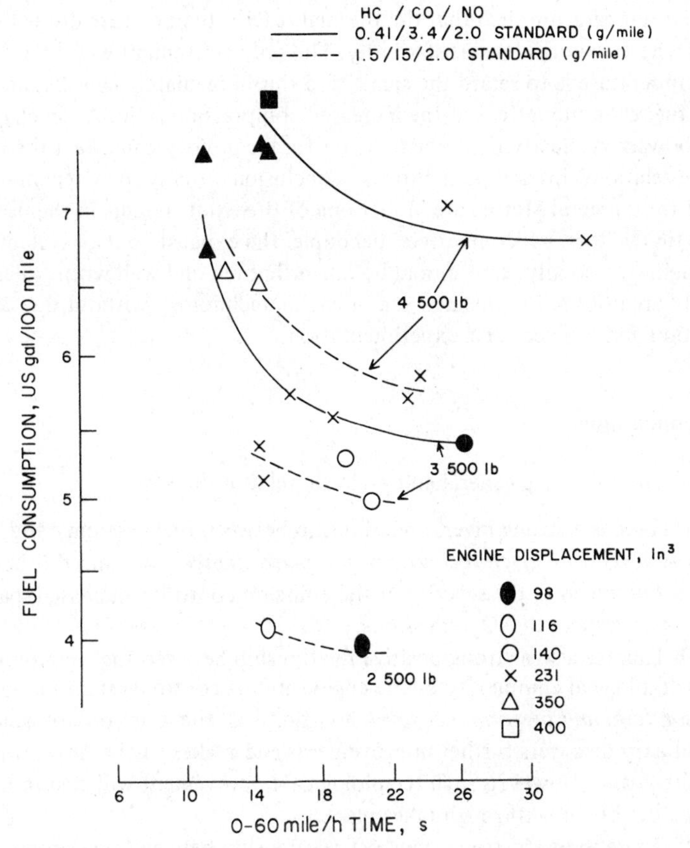

Figure 8.9 Variation of fuel consumption with acceleration performance for cars with different weights and engine sizes at two different emission levels[15] (1 US gal/100 mile = 2.352 l/100 km; 60 mile/h = 96.5 km/h; 1 in^3 = 0.0164 l; 1 lb = 0.454 kg)

derived for a given engine as a consequence of the engine tuning that is shown in a performance map combined with a varying rear-axle ratio.) The figure shows very clearly that the imposition of tighter emission controls gives rise to a poorer fuel consumption–vehicle performance curve, though it is just possible to reduce the engine size by a large enough amount and to trade off enough performance that the fuel economy can almost be held constant. Whether any customer would wish to buy a vehicle with such an engine is a moot point.

There are some rather surprising relationships between emission levels and fuel economy that can occur. Thus Marks and Niepoth[15] show that, for an emission control technology consisting of a thermal reactor only, an increase in compression ratio does not give rise to the expected improvement in fuel economy. This is because increased compression ratio results in a decreased exhaust temperature such that the thermal reactor temperature drops to a point where it works less satisfactorily. The only convenient way to restore the temperature is to retard the spark, and this immediately cancels the beneficial fuel economy effect of the increased compression ratio. As an example, this shows very clearly that emission control technology can affect the usual engine relationships and points to two conclusions: firstly, new technology is called for (General Motors are in fact one of the front-runners in the use of catalytic reactors, which in effect 'decouple' the exhaust control system from the engine); secondly, care should be taken that old and well-worn 'rules-of-thumb' are not used in this new era of engine technology without due care, attention and, if necessary, experimentation.

8.5 Conclusions

(1) The following general and practical relationships exist.

(i) There is a strong inverse relationship between fuel economy and emission level *for a given technology and performance level*, and this becomes more pronounced as the severity of the emission control is increased, particularly with respect to NO_x emissions.

(ii) There is also a strong positive relationship between fuel economy and the technological complexity of the engine and its control system *for a given emission level and performance level*, and this is all the more pronounced as the industry discovers further improvements and makes further inventions over the years. (However, such technological improvement will cost money unless a double breakthrough takes place.)

(iii) There is also a strong trade-off relationship between fuel economy and performance *for a given emission level and technology*, very little different from the traditional one that has been known for many years.

(2) There is no general thermodynamic link between engine efficiency and exhaust emissions, although because an engine is subject to the limitations of, for example, kinetics and surface chemistry, it is not clear how close it will be possible to come to the thermodynamic ideal, and only continued experimental research efforts will tell.

(3) In Europe in the near future two effects will be noticeable.

(i) There will be some loss of overall efficiency as an indirect consequence of the decreasing content of lead in gasoline. If the gasoline octane quality is held constant, this loss of efficiency will take place at the refinery; if the gasoline octane quality falls, the loss will take place in the car because of the reduced compression ratio permitted.

(ii) There will probably be a small gain in fuel economy because of the imposition of relatively mild CO and HC controls, as long as the NO_x control stays constant at its current level.

(4) In Europe in the further future, when and if tighter CO, HC and NO_x controls are introduced, there may be a deterioration in fuel economy, though this is likely to be diminished because the motor industry will have been given sufficient lead time to evolve the appropriate technology.

(5) In the event of extremely stringent emission controls such as have been been suggested in the US and Japan, a radically different engine may be the only way to meet these limits and to preserve good fuel economy. Unfortunately, the NO_x emission level is very critical, for at the 0.4 g/mile level certain of these engine options are simply not possible.

References

1. C. LaPointe. Factors affecting vehicle fuel economy. *Soc. Automot. Eng. Pap.*, No. 730791 (1973)
   Also *Automot. Eng.*, **81** (1973) 46
2. T. C. Austin and K. H. Hellman. Fuel economy of the 1975 models. *Soc. Automot. Eng. Pap.*, No. 740970 (1974)
3. T. C. Austin, R. B. Michael and G. R. Service. Passenger-car fuel economy through 1976. *Soc. Automot. Eng. Pap.*, No. 750957 (1975)
4. *Du Pont Tech. Memo.*, No. Auto-7024 (1974)
5. T. C. Austin and K. H. Hellman. Passenger-car fuel economy – trends and influencing factors. *Soc. Automot. Eng. Pap.*, No. 730790 (1973)
6. J. D. Murrell. Factors affecting automotive fuel economy. *Soc. Automot. Eng. Pap.*, No. 750958 (1975)
7. *Du Pont Pet. Chem. Tech. Conf., 1975*
8. H. C. MacDonald. Fuel consumption trends in today's vehicles. *Soc. Automot. Eng. Pap.*, No. 730517 (1973)
   Also in *Soc. Automot. Eng. Spec. Publ.*, No. SP-383 (1973)
9. J. G. Huebner and D. J. Gasser. Energy and the automobile – general factors affecting vehicle fuel consumption. *Soc. Automot. Eng. Pap.*, No. 730578 (1973)
   Also in *Soc. Automot. Eng. Spec. Publ.*, No. SP-383 (1973)
10. *Fuel Economy and Emission Control.* Environmental Protection Agency Mobile Source Pollution Control Program, Washington (November 1972)
11. H. Toulmin. Private communication, cited in reference 1
    See also chapter 3 of this book
12. K. Oblander and B. May. Design and special development problems of Mercedes–Benz V8 engines. *Soc. Automot. Eng. Pap.*, No. 750051 (1975)
13. J. H. Boddy, P. E. Bright and G. S. Parkinson. Air pollution in Europe and the fuel shortage. Paper presented at *Inst. Mech. Eng.–Soc. Motor Manuf. Traders Conf., 1974*, p. 123
14. *Motor Vehicle Emissions.* US National Academy of Sciences, Washington (November 1974)

15. C. Marks and G. Niepoth. Car design for economy and emissions. *Soc. Automot. Eng. Pap.*, No. 750954 (1975)
16. J. J. Gumbleton, R. A. Bolton and H. W. Lang. Optimizing engine parameters with exhaust gas recirculation. *Soc. Automot. Eng. Pap.*, No. 740104 (1974)
17. *Automot. Eng.*, **83** (October 1975) 44
18. H. W. Barnes-Moss and W. M. Scott. The small high-speed diesel engine – a power unit for future passenger cars. *Ricardo Consult. Eng., Shoreham-by-Sea, Sussex, Pap.*, No. DP 17478A (February 1974)

9 The Measurement of Fuel Economy

R. BURT

9.1 Introduction

It is a truism to state that the study of the effect of fuel and lubricant properties on gasoline consumption must be based on adequate techniques for the measurement of consumption (or its reciprocal expression, fuel economy). However, although this is self-evident, it is unfortunately also true that outside the US there are no widely accepted methods for the realistic assessment of fuel consumption by vehicles, and it is only since 1973 that the US has been in advance of the rest of the world.

The reality of the situation is highlighted by the way motor manufacturers use fuel consumption data in their promotional material. The only available data are those published by the technical motoring press, either the so-called overall consumption, or values measured at constant speed. The overall value depends so much on the usage of the particular vehicle, which is clearly variable, that the significance of constant-speed data in relation to normal road use is arguable. The misuse of these types of data has recently been criticized by the technical motoring press[1] : however, at the same time it is pointed out that no satisfactory alternative exists at present.

Before discussing the relative merits of different techniques for the measurement of fuel economy, it is germane to consider the factors which influence the fuel consumption of a particular passenger car. These are listed in table 9.1. The variables considered are those which relate to vehicle usage; factors such as vehicle weight, tyre type and fuel and lubricant effects are not considered, and it is assumed that the car meets the manufacturers' specification.

It is manifest that all these factors will be influential, and their effect can be illustrated by the results of a simple test on a traffic-free closed road circuit (the Thornton Research Centre (TRC) Appleton test track), where consumption was doubled by increasing the average speed in non-stop operation from 56 to 100 km/h.

The magnitude of the measurement problem posed by usage variables can be judged against the potential effects of variations in fuel or lubricant properties. This subject has been discussed in detail elsewhere in this book,

Table 9.1 Vehicle usage factors influencing fuel consumption

Driving duty	Urban, suburban, highway, motorway
Traffic conditions	Degree of congestion on road
Driving habit	Average speed, rate of acceleration and deceleration, manual choke operation
Ambient conditions	Temperature, wind, rain

but it is generally considered that individual fuel or lubricant effects on consumption are less than 5%.

It is clear that the most precise* measurement of consumption will be obtained under the most closely controlled conditions, where all the factors in table 9.1 are defined (or are eliminated). However, this can lead to a situation where the significance of the measurement is remote from practical vehicle operation; indeed, there is a potential conflict between the degree of control and the relevance of the measurement.

The types of test for consumption measurement which will be considered are as follows.

(a) Uncontrolled road tests.
(b) Controlled road tests.
(c) Cycle tests on the road (including constant-speed tests).
(d) Cycle tests on chassis dynamometer (including constant-speed tests).
(e) Bench engine tests.

As well as the type of test, the equipment used for the determination of the quantity of fuel consumed will also be described and discussed.

9.2 Measurement in Uncontrolled Road Tests

An uncontrolled road test can be defined as one in which none of the usage variables given in table 9.1 is controlled. The selection of the car, its maintenance and the fuel used may be defined, but the driving pattern and driving habit, etc., are those normally experienced in normal road use.

This is the most difficult type of test in which to measure consumption, though it is the most relevant to the market-place. It is not difficult to measure the amount of fuel consumed over a considerable distance, and the accuracy of metering of a kerbside pump (to better than 1%) is adequate for this purpose. The calibration errors of the odometer are a more serious

*For a definition of the statistical terms commonly used in connection with fuel economy measurement, see appendix B.

problem, and it is desirable to use a specially calibrated instrument or, at least, to calibrate the instrument fitted to the car.

9.2.1 Reproducibility

The major problems in tests of this type arise from the lack of control over the usage variables. This can be illustrated by the results of TRC fuel consumption measurements on a set of eight pairs of different employee-owned cars. The fuel consumption was determined at 6000 mile intervals and the rate of mileage build-up was at least 12 000 miles per year. The differences in consumption of the paired cars ranged from 0.3 to 15.1% with a mean of 7.0%.

In another test, with six examples each of three different cars, the spread in consumption for one model varied from 11 to 32%. In this test the consumption was measured over 10 000 miles.

It is noticeable that the variation for a vehicle of one type due to driving pattern and driving habit is much larger with a six-car sample (mean spread, 20%) than with the paired cars (mean difference, 7%). This is inherent in an uncontrolled test and relates to the probability of combinations of driving habit and driving duty which gives extremes of fuel consumption. The differences in repeat tests with the same car, driver and fuel varied from 1.7 to 14.4% with a mean of 5.2%.

In addition to changes in usage pattern for one driver in a particular car (which will be smaller than those for different drivers in the same car model) there are two factors which influence the reproducibility: the deterioration in engine condition over a relatively long period; differences in mean ambient conditions over the test periods (e.g. tests in summer or winter). The latter can be minimized by using either short or very long periods of measurement; however, a short period (and consequently low mileage) gives inaccurate results, and a long period aggravates the wear factor.

9.2.2 Fuel Supplies

It is considered that the test duration to obtain a significant result in an uncontrolled test should be at least 1000 miles (1600 km). The quantity of fuel consumed over this distance, about 33 gal (150 l), can create problems of expense and availability when special fuel blends are used. In longer uncontrolled tests with large numbers of vehicles the very large quantities of fuels consumed can present problems of quality control if successive batches have to be prepared.

9.2.3 Cost

The difficulty of obtaining significant results in fuel (or lubricant) tests is highlighted by the relationship between test repeatability and fleet size. With a standard deviation of the samples of 7.5%, fleets of 46, 81 and 182 cars would be required to establish economy differences of 4%, 3% and 2%, respectively. The cost and duration of such a test makes this technique unattractive, despite the fact that the end result would be directly related to normal vehicle use and thus would be generally acceptable, for example, to the public.

9.2.4 Accuracy

The accuracy of fuel consumption measurement in an uncontrolled road test can be substantially improved by the use of the 'trip-mileage analyser'[2]. This instrument was developed at TRC and records engine revolutions and time in bands of engine load operation defined by manifold vacuum, allowing tests to be normalized to a common driving pattern. With this equipment the measured consumption of fuel in 24 fully warmed-up trips with one car on the same fuel has been correlated with recorded trip data with a standard error of 2.2%. The trips used varied from congested urban routes at an average speed of 5.9 km/h to high-speed motorway driving at an average speed of 129 km/h. The fuel economy ranged from 10.3 to 36.5 mile/gal (from 27.4 to 7.74 l/100 km).

In its present form, however, the trip-mileage analyser cannot be used in cold-start journeys which include choke operation. A more serious defect is that it is not (as presently developed) suitable for the identification of fuel effects on consumption because gasoline quality, e.g. gravity or volatility, can change the relationship between manifold vacuum and engine load, which is used implicitly in the normalizing procedure. Manifold vacuum thus ceases to be a unique descriptor of engine load when different fuels are used. Measurement of engine torque would be more reliable in this respect.

9.3 Measurement in Controlled Road Tests

Before discussing the application of controlled road tests in the measurement of fuel consumption it is necessary to define what is meant by 'controlled'.

A controlled test is one in which one or more of the variables in table 9.1 are held constant. The degree of control can obviously vary widely from the simplest case, with cars driven over a similar route during a prescribed time period, to the highly controlled case, where matched cars are driven in convoy operation over a strictly defined route. (We do not here include cycle tests on the road; these are discussed in a later section.)

In the first example there will be a moderate degree of control over the driving duty, depending on whether all the cars are driven along the same route or along different routes of a generally similar kind. The driving habit will be controlled for each car if the car–driver combination is fixed but only to the extent that there is consistency in one person's driving habits. The variation in the traffic condition is only defined by the degree of consistency of road congestion over a particular route during a prescribed time period; it is inevitable that there will be a substantial residual variation in traffic conditions. Ambient conditions cannot be controlled, and variation can only be minimized by limiting the duration of the test to a relatively short period. Even this does not ensure consistency in a highly variable weather situation, as in the UK.

Work carried out at TRC is an example of a lightly controlled road test using employee-owned cars on long and short trips. The former were long-distance weekend runs over the same general route for each car, and the latter were weekday driving to and from work over the same route for each car. The long-trip test covered two summer months, and the short-trip test three winter months. It was concluded that a 2% difference in fuel consumption could be detected in trips of about 400 km. The repeatability was naturally poorer in the series of short trips, of about 20 km each, and differences of about 5% could be detected in tests on a Monday–Friday basis, i.e. over a total distance of about 200 km. In considering such precision in relation to potential fuel effects it is clear that more testing will be required to quantify fuel volatility effects, which may influence economy in cold-start short trips, than to quantify gravity effects, which will influence economy in both long and short trips.

The repeatability of measurements under more controlled conditions, in a single car with the same driver, has been studied at TRC to determine differences in fuel economy with gasoline, butane and liquefied natural gas (LNG). The car was driven over two fixed routes, one simulating an inter-city journey, the other a taxi route. The residual standard errors corrected to equivalent distances of 1000 km are given in table 9.2.

The measurement of fuel economy for all fuels was about twice as accurate for the taxi route as for the inter-city route; it was considered that this was due to the greater control over the driving pattern in the shorter taxi route.

Table 9.2 *The repeatability of fuel consumption measurements over fixed road routes*

Route	Fuel	Residual % standard error of fuel consumption
Inter-city	gasoline	4.4
	butane	1.9
	LNG	1.9
Taxi	gasoline	2.2
	butane	0.95
	LNG	1.0

In both routes the errors with gasoline were about twice those with gaseous fuels. The effect on the carburetter of uncontrolled operation of the choke and the accelerator pump operating on gasoline, and the use of volume measurement (fuel tank filling) in place of weighing (as with the gaseous fuels), were thought to contribute to the reduced repeatability with gasoline operation. These results illustrate some of the problems which emerge in the measurement of fuel consumption as precision is improved by increasing the degree of test control. Factors which are apparently insignificant become important as the accuracy of measurement increases.

An indication of the precision that can be attained in a highly controlled road test is given by the results from a short trial on a closed road circuit (TRC Appleton test track). Three cars were used and driven in top gear around the circuit in a gentle manner at an average speed of 58 km/h. Fuel consumption was measured with a calibrated Petrometa and distance with a fifth wheel. The tests on each car were carried out in succession. The results are given in table 9.3.

A test of this type borders on a cycle test; however, there was no positive control over the pattern of driving as in a prescribed cycle. As would be expected on a closed circuit, the distance repeatability is very good, reflecting the ability to follow the same path each time around. The repeatability of the fuel consumed reflects the size of the engine: the precision improves as the amount consumed increases.

Convoy operation with a group of cars with interchange of drivers and car position represents the ultimate in a controlled road test. This technique permits a number of cars of either the same or different type to be compared in conditions where all the factors in table 9.1 are identical. However, this only applies for a single test, since normally it will be impossible to repeat some of the conditions on another occasion (e.g. traffic congestion and ambient conditions will vary from day to day). This substantially reduces the advantage of this technique, which is also very expensive in terms of manpower.

At this point it is necessary to introduce a philosophical concept relating to fuel economy measurement. It is clearly important to consider whether or

Table 9.3 *The repeatability of distance travelled and fuel consumed on a closed road circuit at an average speed of 58 km/h*

Car	Engine size, cm^3	No. of tests	% standard deviation	
			Distance	Fuel consumed
Fiat 127	903	4	0.03	2.4
Chrysler Avenger	1600	8	0.1	1.5
Ford Consul	2500	4	0.08	0.93

not economy comparisons should be made under conditions where total vehicle operation is strictly defined. For example, in convoy operation, where vehicle road duty is inherently the same for all the vehicles, there can still be a number of driver variables which will influence economy. Gear-change points may not be the same, and during warm-up the patterns of choke release will vary. In a strictly defined test these driver variables would be specified in the test procedure, but this does not necessarily reflect the pattern of normal vehicle usage. For instance, a more volatile fuel may permit a manual choke to be released earlier, thus improving economy compared with a less volatile fuel. Choke operation has been looked upon (in table 9.1) as an aspect of driving habit, and in a completely controlled test will itself be defined, thus eliminating (at least partially) the beneficial effect of the more volatile fuel on consumption. A fuel factor such as volatility influences only one of the usage factors which may be held constant in a test but which may vary in practice. There are many ways in which a driver may interact with the vehicle and thus may influence the way in which the vehicle is driven. For example, cars with good road-holding will generally be driven faster than those which handle badly, and a driver will be forced to change to low gears at higher speeds with cars with poor engine flexibility.

9.4 Measurement in Cycle Tests with Cars Driven on the Road

The dividing line between a fully controlled road test and cycle tests on the road is, to a large extent, arbitrary. A cycle test is defined here as a test in which a vehicle is driven to a prescribed pattern of speed and time. The cycle includes the points at which gears are changed (in a manual transmission) and at which the brakes are applied. This definition will cover measurements made at idle and at steady-state speeds, conditions which can be considered as the simplest cycles.

9.4.1 Road Test Cycles

9.4.1.1 Manufacturers' Tests

Traditionally, motor manufacturers and other bodies have used cycles for the measurement of fuel consumption as an essential part of vehicle development programmes. Inevitably, each manufacturer would develop a cycle which is considered appropriate for his use, and the complexity of the resulting situation can be judged from table 9.4 (taken from reference 3) which lists the cycles used in the US.

These cycles were designed to include all driving duties, as defined in table 9.1, but it is obvious from table 9.4 that the manufacturers' view of the make-up of a particular driving duty can vary widely. For example, in

Table 9.4 Comparison of US fuel economy road cycles as developed by the motor manufacturers

No.	Economy cycle	Average speed, mile/h	Cycle distance, mile	Stops per mile	Idle time, s/mile	Acceleration rate, ft/s^2	Deceleration rate, ft/s^2	Road-load speeds, mile/h
1	Ford city	15.6	3.6	5.6	69[a]	5,7	7	25,30
2	General Motors business	16	2.0	4.0	30	5,8[b]	5,6	15,20, 25,30
3	Chrysler urban	16.7	4.6	5.2	36	3	5	26,35
4	General Motors suburban	24	3.7	1.6	20	3,6,8,WOT	3,6,CT	25,30, 35,40
5	Proposed Chrysler suburban	32.4	5.2	0.4	13	2,4	2,4	30,40, 50
6	Ford suburban	42	5.2	0.4	1.2	3,5,7	5,10	40,50, 60
7	General Motors highway	47	14.8	0.3	2.0	2,5,WOT	4,6,CT	50,55, 60,70
8	Chrysler inter-state 50	50	4.7	0	0	1	1	50
9	Chrysler inter-state 70	70	4.7	0	0	1	1	70
10	General Motors inter-state	70	14.9	0.1	0	2,3,6	6,CT	60,70, 75

WOT, wide-open throttle; CT, closed throttle.

[a]Includes 21 s/mile of U-turns.

[b]General Motors' acceleration rates are initial rates with constant throttle acceleration to road-load speed.

urban cycles average speeds varied from 24 to 42 mile/h, maximum speeds from 40 to 60 mile/h and number of stops per mile from 0.4 to 1.6.

9.4.1.2 The Involvement of the Society of Automotive Engineers

While the individual manufacturers had their own cycles, the situation in the US did not accord with the need for standardized cycles for fuel economy measurement. Consequently, in 1973 the Environmental Protection Agency (EPA) commissioned the Society of Automotive Engineers (SAE) to develop test cycles for the measurement of fuel economy in different driving duties. At this time, fuel economy under urban driving conditions was already being measured by the existing emission test cycle[4,5] (the FTP-LA-4 cycle run on a chassis dynamometer; this procedure will be discussed in the next section).

The SAE task force decided that the LA-4 cycle, though representative of urban driving, was too complex to be used on a road track and that constant-speed tests did 'not necessarily relate to conditions normally encountered by consumers', a point to be discussed later. It was then decided to adopt the cycles numbered 2, 6 and 9 in table 9.4.

The reproducibility of fuel economy measurements in the test cycles was assessed in two exchange schemes with two different sets of three car models, ranging in engine size from 2 to 5.6 l. The results are shown in table 9.5.

The results did not indicate any significant differences in precision between the test cycles, but they did show that on average the reproducibility in the second series was twice as good as that in the first series, probably indicating improvement with practice.

In addition to the urban, suburban and inter-state (motorway) cycles in table 9.5, the SAE adopted a second inter-state cycle with an average speed of 55 mile/h to match the US federal speed limits and to be more compatible with chassis dynamometer testing. It has also been suggested by the SAE that the adoption of two speeds in similar inter-state cycles would demonstrate the benefits to the motorist of reducing speeds. (This is an example of the

Table 9.5 Reproducibility of fuel economy measurements in road test cycles adopted by the Society of Automotive Engineers

Test cycle	% standard deviation of measurements	
	Test 1	Test 2
General Motors business	3.11	1.76
Ford suburban	3.23	1.23
Chrysler inter-state 70	3.65	1.40

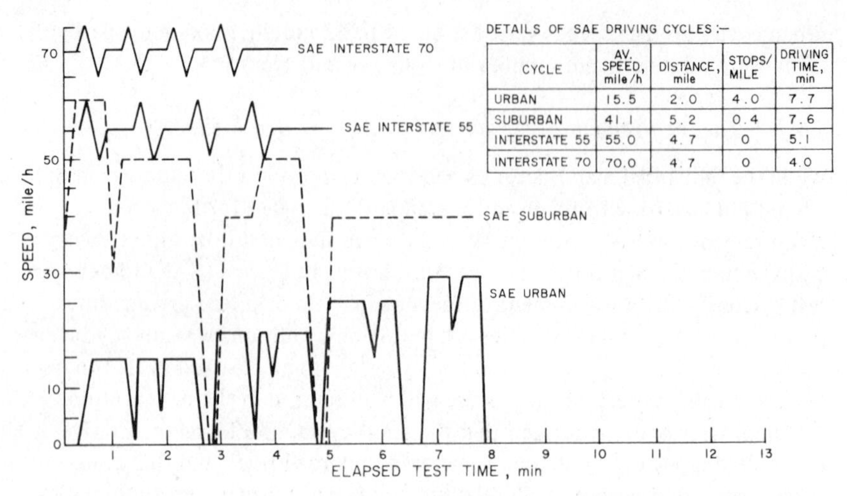

CYCLE	AV. SPEED, mile/h	DISTANCE, mile	STOPS/ MILE	DRIVING TIME, min
URBAN	15.5	2.0	4.0	7.7
SUBURBAN	41.1	5.2	0.4	7.6
INTERSTATE 55	55.0	4.7	0	5.1
INTERSTATE 70	70.0	4.7	0	4.0

Figure 9.1 Trace of speed versus time for the three SAE road cycles (1 mile/h = 1.609 km/h)

potential socio-economic application of fuel economy measurements in test cycles.)

The patterns of the four SAE cycles are illustrated in figure 9.1. The cars on these cycles are all driven fully warmed up and therefore cannot demonstrate the effects of cold-start and vehicle warm-up. The use of a cold-start procedure was deferred by the SAE since it would introduce greater variability into the measurement and would render replication very time consuming because of the extended soak time (12–24 h).

It will be noted in figure 9.1 that all four SAE cycles are relatively simple in pattern. This is an important feature of a test cycle to be followed on the road, and complex cycles such as the FTP-LA-4 emission test cycle are unsuitable for road use. A complex cycle can only be followed with a visual driving aid, as used to drive chassis dynamometer cycles. (This instrument will be described in a later section.) Road cycles are followed on a marked test track or with the help of a tape recorder; both these techniques can be used for cycles of the type shown in figure 9.1.

9.4.1.3 Road Tests at the Thornton Research Centre

Tests at TRC with a simple road cycle have shown that the repeatability of consumption measurements can be excellent, thus confirming the SAE results given in table 9.5. The car was driven fully warmed up, and the cycle started with a 30 s idle period which was followed by acceleration through the gears up to 30 mile/h in top gear. A tape recorder was used to control this part of the cycle. A speed of 30 mile/h was maintained until a marker indicated the start of deceleration to a standstill. Details are given in table 9.6.

Table 9.6 Simple road cycle used to measure fuel consumption at the Thornton Research Centre Appleton test track

Average speed,	mile/h (km/h)	23.3 (37.5)
Maximum speed,	mile/h (km/h)	30 (48.3)
Cycle distance,	mile (km)	1.94 (3.12)
Cycle time,	min	5
Stops per mile		0.5
Idle time,	min	0.5

Tests with 27 cars ranging in engine size from 600 cm^3 to 2 l showed a percentage standard deviation ranging from 0.35 to 1.48 with a mean of 0.93. This precision is better than that obtained in the SAE cycles (table 9.5), but the cycle used is also less complex.

9.4.1.4 Steady-speed Road Tests

The measurement of fuel consumption at steady speed is a form of highly controlled road test. Measurements obtained by this method have been reported for many years in car tests carried out by the technical motoring press.

Recently the French government published[6] a method for the measurement of fuel consumption which included measurements at two speeds, 90 and 120 km/h. This procedure has been adopted in the proposals by the Groupement des Rapporteurs de la Pollution de l'Air (GRPA)[7] for a European test method for the measurement of vehicle fuel consumption.

The method specifies that four tests shall be made, two at an average speed above the reference speed and two below the reference speed. The length of the test run has to be at least 2 km. The speed for each test must be kept steady within ±2 km/h, and the average must not differ from the reference speed by more than 2 km/h. The fuel consumption at the reference speed is determined by graphical interpolation from the test results. The foregoing illustrates the relative complexity of the simplest fuel consumption measurement which can be determined in a controlled road test.

There is no information at present available on the precision of measurement obtained using this procedure.

Variations in the ambient conditions and the levelness of the road or test track are two factors which can significantly alter the fuel consumption

Table 9.7 Ambient conditions and road quality specified in road-test procedures of the Society of Automotive Engineers and of the Groupement des Rapporteurs de la Pollution de l'Air

Requirements		SAE[8]	GRPA[7]
Ambient conditions			
Temperature,	°C	−1 to 32	5 to 25
Pressure,	mm Hg	–	730–756
Relative humidity,	%	–	<95
Maximum wind velocity,	km/h		
steady		24	11
gusts		32	30
Road conditions			
Difference in level,	m	–	<4
Gradient,	%	<½	<2

measured in a controlled road test. The tolerable variations have to be specified, and those laid down by the SAE and the GRPA are given in table 9.7.

The variations in table 9.7 are relatively large; however, very narrow limits on ambient or road conditions would severely limit the number of days or the roads and test tracks on which measurements could be made.

The SAE procedure includes an elaborate correction equation to take account of ambient and fuel temperature, ambient pressure and fuel gravity in the calculation of the corrected fuel consumption. The equation is

$$\text{corrected mile/gal} = \text{observed mile/gal}\,(T_S\text{CF})(P_B\text{CF})(\text{SG}_F\text{CF})(T_F\text{CF})$$

where CF is the correction factor, T_S the average ambient temperature, °F, T_F the average fuel temperature at measuring instrument, °F, P_B the average barometric pressure, in Hg, and SG_F the specific gravity of the fuel at 60°F.

$$T_S\text{CF} = 1 + 0.0014(60 - T_S)$$

$$P_B\text{CF} = 1.0 \text{ for urban cycle}$$

$$= 1.0 + 0.0072(P_B - 29.0) \text{ for suburban cycle}$$

$$= 1.0 + 0.0084(P_B - 29.0) \text{ for inter-state 55}$$

$$= 1.0 + 0.0144(P_B - 29.0) \text{ for inter-state 70}$$

$$T_F\text{CF} = \frac{1}{\text{multiplier for volume reduction to 60 °F*}}$$

$$\text{SG}_F\text{CF} = 1 + 0.8\,(0.737 - \text{SG}_F)$$

*Obtained from charts or tables[8].

These factors correct to standard conditions of 60 °F (15 °C) air and fuel temperature, 29.0 in Hg barometric pressure and 0.737 fuel specific gravity. This correction equation is a composite of the empirically derived correction techniques of US manufacturers.

The GRPA[7] method gives corrections for fuel specific gravity and temperature but only specifies, without correction procedures, that the atmospheric density must not differ by more than 5% from that at the reference conditions of 1000 mbar and 25 °C.

9.4.2 Measurement of Fuel Consumed

The quantity of fuel consumed in a controlled road test is normally measured by a volumetric technique. Fuel weighing can be used for cycles with a standing start and stop (e.g. the SAE urban and suburban cycles), but it is not convenient, and the potential marginal improvement in precision is not practically significant.

The volume of fuel consumed can be measured with a flow meter or a burette system. Flow meters commonly used include diaphragm meters (e.g. Petrometa), piston displacement meters (e.g. Tokico) and more recently the Pierburg Luftfahrtgehrate Union GmBH (PLU) gear-type meter. These instruments will be described in a later section.

The flow meter is normally fitted between the vehicle fuel pump and the carburetter; thus a pressure drop across the meter can reduce the pressure of fuel delivery to the carburettor. The SAE procedure emphasizes the desirability of keeping the pressure drop as small as possible to avoid influencing carburetter performance. Excessive pressure drop at high fuel flows can be a problem with some piston-type flow meters.

Burette systems are generally installed so as to feed the vehicle fuel pump, and this will not affect fuel pressure; however, the burette then has to be vented, which is undesirable and can lead to vapour loss. A burette which can be placed between the pump and the carburetter has been developed at TRC. It does not alter fuel pressure and is sealed from the atmosphere. This burette was used in the tests described earlier (see table 9.6).

The accuracy required of the fuel-measuring system is defined in both the SAE and GRPA test methods. SAE specify that the device must be capable of measurement in increments of less than 4 ml and must be accurate to ±0.5% of the indicated consumption. In the 2.0 mile SAE urban cycle a sub-compact car would only consume about 400 ml of fuel: with 4 ml increments this would give an accuracy of 1%, at best. The GRPA proposals specify an accuracy of ±2%. It is doubtful whether existing fuel flow meters can consistently attain an accuracy of 0.5%, and a figure of 1% is more realistic. The absolute accuracy of the measurement device is important in the measurement of absolute consumption. Good repeatability of the measurement is adequate for tests designed to identify fuel effects.

The controlled road test for the measurement of fuel consumption can clearly give a precision capable of detecting and quantifying fuel effects. The standard deviation of measurement ranges from approximately 1% for simple cycles to 2% for more complex cycles. In addition, the test result is meaningful in so far as the test cycles are representative of a particular mode of vehicle application. However, the utility of this type of measurement is limited by the lack of control over ambient conditions. This is a particular problem in countries with variable weather, such as the UK. The need for a high-quality test track with small gradients and long straight sections suitable for high-speed driving is also a limitation.

9.5 Measurement in Cycle Tests with Cars Driven on a Chassis Dynamometer

The use of a chassis dynamometer brings a new dimension into the measurement of fuel economy, for with a temperature-controlled chassis dynamometer all the usage factors in table 9.1 can be controlled, or their effects can be eliminated.

In a test of this type, fuel consumption is measured with the vehicle driven on chassis dynamometer rolls either to a test cycle followed from a driver's aid or at constant speed. The dynamometer should be designed to impose the correct inertia forces and road loads onto the vehicle, so that operation on the road is simulated. It is clear that, in principle, chassis dynamometer tests offer powerful advantages over a test carried out on the road; in practice, however, they have their special problems.

9.5.1 Chassis Dynamometer Test Cycles

A chassis dynamometer test was used in the first application of a standardized test for the measurement of fuel economy. This was the proposal by the EPA in the US to use the federal emission test cycle as the basis for the fuel economy labelling of US cars. This 7.5 mile urban dynamometer driving schedule[5] (UDDS) was based on the so-called 'LA-4 route', a 12 mile road route in central Los Angeles and is illustrated in figure 9.2(a). The use of this cycle was logical in that it is convenient to use the same cycle for the measurement of fuel economy and exhaust emissions. In addition, US motor manufacturers and others had a great deal of experience with this cycle.

It is desirable that a cycle used for the measurement of fuel economy should be representative of a particular driving pattern, and studies have been carried out to compare the UDDS with urban driving in a number of US cities[5,9]. Some of the results are given in table 9.8.

The data in table 9.8 show that the UDDS is representative of driving in central Los Angeles but that average speeds are lower and that idle times higher than those in other cities. The status of the UDDS in the US for emission control ensured its subsequent adoption as a measure of urban fuel consumption, and considerable data have been published on the measured fuel consumption of 1973[10], 1974[11] and 1975[12] model year cars.

The fuel economy measured in the cold-start UDDS is poor, as might be expected from the nature of the cycle. For example, the mean consumption for US 1973 cars with an equivalent inertia weight of 2500 lb (1134 kg) (similar to European medium saloon cars) is 19 mile/US gal (12.4 1/100 km). It was suggested by the US motor manufacturers that the economy measured in this test was worse than average for a car in normal usage since the UDDS

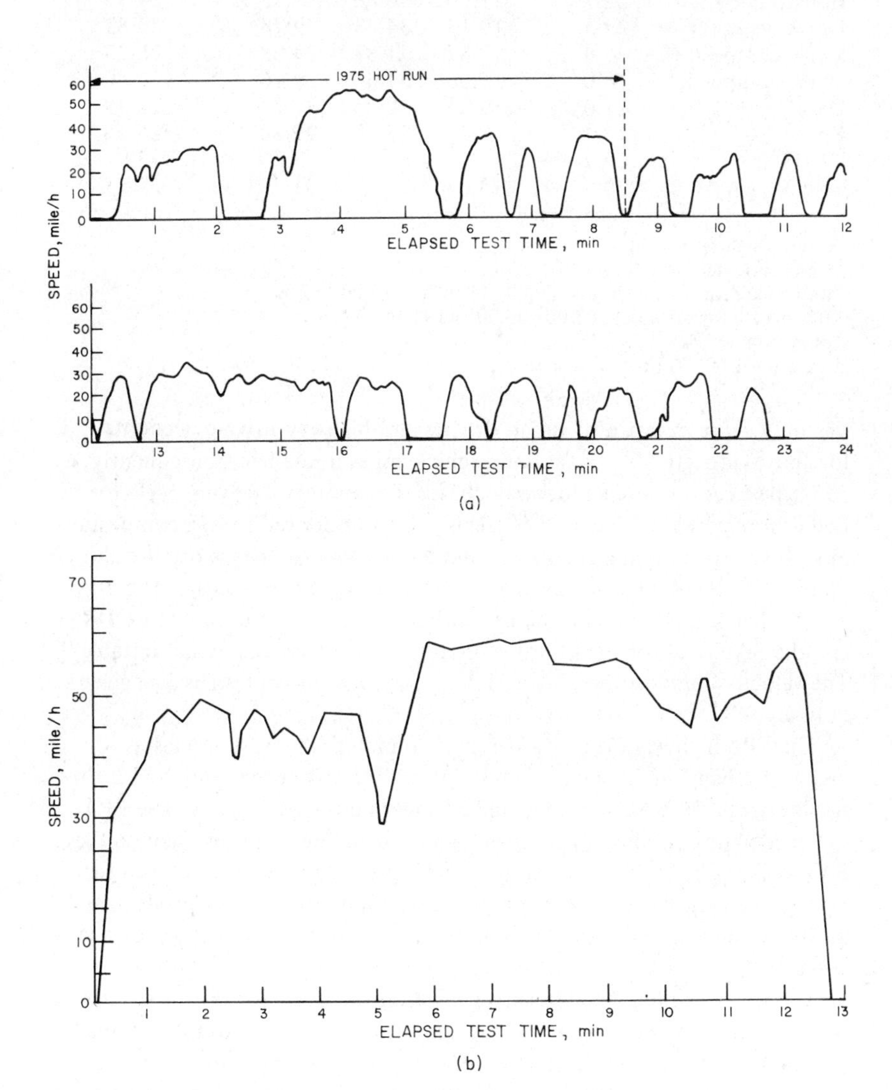

Figure 9.2 (a) Trace of speed versus time for the EPA UDDS cycle. (b) Trace of speed versus time for the EPA highway cycle (1 mile/h = 1.609 km/h)

Table 9.8 Comparison of urban driving patterns and driving cycles

Cycle or city	Average speed, mile/h	Percentage of time in each mode			
		Idle	Cruise	Acceleration	Deceleration
New York City	21.6	17.45	26.49	29.12	26.95
Chicago	24.5	14.11	30.86	28.30	26.78
Cincinnati	25.9	11.34	30.72	30.89	27.06
Houston	27.7	11.30	36.80	27.35	24.58
Los Angeles	29.3	10.13	34.28	29.78	25.82
5-city composite[a]	25.8	12.87	31.83	29.08	26.23
5-city composite[b]	26.0	13.06	31.50	29.16	26.30
LA-4[c]	21.0	13.56	27.25	31.73	27.49
LA-4[d]	17.4	18.43	25.42	29.82	26.28
UDDS[e]	19.7	18.2	30.2	27.7	23.9
Original LA-4[e]	20.9	13.6	27.3	31.7	27.5

[a]Cities weighted equally.
[b]Cities weighted by vehicle registration.
[c]Results are for defined hours: 09.00–11.00 and 13.00–15.00.
[d]Results are for off-hours: 07.00–09.00 and 15.00–17.00.
[e]Data from ref. 5.
1 mile/h = 1.609 km/h.

was only representative of urban driving and highway driving accounts for approximately 50% of the total vehicle miles in the US. Consequently, in 1974 the EPA proposed a highway chassis dynamometer driving cycle for fuel economy measurement[13,14]. This cycle was derived from a composite of highway road driving and has maximum and average speeds of 59.9 and 48.8 mile/h (96.4 and 78.5 km/h) respectively (see figure 9.2(b)). The highway cycle is compared with the EPA urban economy test using the UDDS and the SAE road (or track) test economy cycles, discussed earlier in table 9.9. The velocity composition of the EPA highway and urban tests is also given in figure 9.3.

The EPA highway cycle was designed to incorporate various classes of highway driving, including freeways with a 55 mile/h speed limit, and it thus falls between the SAE suburban and 55 mile/h inter-state cycles. The EPA cycles include considerably more acceleration and deceleration than do the SAE cycles. This is partly because the SAE procedures were developed for road (or test track) use where complex cycles are difficult to reproduce and partly because the EPA road data indicate that drivers do not travel at steady speeds.

The trend to adopt the legislative emission test cycle for the measurement of fuel economy under urban driving conditions has spread to Europe and is proposed in Japan. The Economic Commission for Europe (ECE) 15 emission test cycle has been proposed for this purpose by the French government[6] and has recently been embodied into a draft ECE recommendation by the

Comparison of current and proposed fuel economy driving cycles

Driving cycle		EPA			SAE			
		CVS-C	CVS-CH	Highway	Urban	Suburban	Inter-state (55 mile/h) (88.5 km/h)	Inter-state (70 mile/h) (112.7 km/h)
Start		Cold (1972-FTP)	Cold (1975-FTP)	Hot	Hot	Hot	Hot	Hot
Test location		Chassis rolls	Chassis rolls	Chassis rolls	Track	Track	Track	Track
Length,	mile (km)	7.45 (11.99)	11.04 (17.77)	10.25 (16.50)	2.0 (3.22)	5.2 (8.37)	4.7 (7.56)	4.7 (7.56)
Driving time,	min	22.87	31.3	12.7	7.7	7.6	5.1	4.0
Average speed,	mile/h (km/h)	19.5 (31.4)	21.18 (34.1)	48.8 (78.5)	15.5 (24.9)	41.1 (66.1)	55.3 (89.0)	70.5 (113.5)
Maximum speed,	mile/h	56.5	56.5	59.9	30	60	60	75
Maximum acceleration,	ft/s^2 (m/s^2)	4.84 (1.48)	4.84 (1.48)	4.69 (1.43)	7.0 (2.13)	7.0 (2.13)	1.0 (0.30)	1.0 (0.30)
Cruise time,	%	7.9	7.7	16.5	58.3	75.2	61.8	51.5
Acceleration time,	%	39.6	39.3	44.4	11.3	11.3	19.1	24.3
Deceleration time,	%	34.6	34.9	38.7	17.4	10.5	19.1	24.2
Idle time,	%	17.8	18.1	0.4	13.0	3.0	0	0
Stops per mile		2.3	2.0	0.1	4.0	0.4	0	0

N.B. Accelerations and decelerations are defined as changes in velocity greater than 0.1 ft/s^2.
CVS-C, cold-start constant-volume sampling test with 1972 UDDS cycle; CVS-CH, cold-start constant-volume sampling test with 1972 UDDS cycle but with first 505 s of cycle repeated.

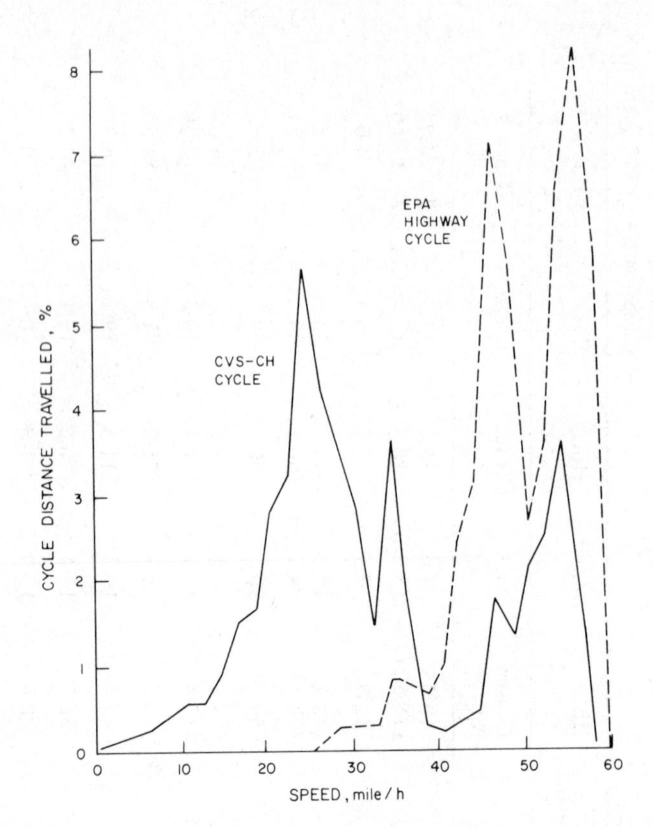

Figure 9.3 EPA urban and highway fuel economy driving cycle velocity distributions[19] (1 mile/h = 1.609 km/h)

GRPA[7]. The British government has proposed the introduction of this test procedure for the measurement of passenger-car fuel economy by the end of 1977. In these proposals cars on the ECE 15 emission test cycle are driven from a fully warmed-up start, in contrast with the cold start in the emission test. This apart, the cycle procedure in terms of driving habit is identical. The Japanese government is proposing the application of the ten-mode hot-start emission test cycle for the measurement of vehicle fuel economy. Details of the ECE 15 and Japanese ten-mode cycles are given in table 9.10, and the patterns of the cycles are shown in figure 9.4.

In comparison with the US UDDS, these cycles are low-severity light-duty cycles (cf. tables 9.9 and 9.10), though they are considered to be representative of the patterns of urban driving in European and Japanese cities.

It is unfortunate that the current European proposals for the measurement of fuel economy do not include a highway type of driving cycle. The proposed use of constant-speed measurements at relatively high speeds is no

Table 9.10 Characteristics of the proposed European and Japanese urban economy test cycles

Feature		Driving cycle	
		ECE 15	Japanese ten mode
Length,	km	1.013[a]	0.664
Driving time,	min	3.25[a]	2.25
Average speed,	km/h	18.7	17.7
Maximum speed,	km/h	50	40
Maximum acceleration,	m/s^2	1.04	0.8
Acceleration time,	%	21.5	24.4
Cruise time,	%	29.2	23.7
Deceleration time,	%	18.5	25.2
Idle time,	%	30.8	26.7
Stops per kilometer		3	3

[a] In the GRPA proposal for a European fuel economy test, consumption is measured over at least three pairs of cycles.

substitute for a realistic cycle test, since it fails to take into account a number of factors present in road usage. These deficiencies are as follows.

(1) The elimination of the effect of vehicle inertia weight, with consequent overemphasis of aerodynamic drag.

(2) The incomplete coverage of the highway driving ranges of speeds and loads.

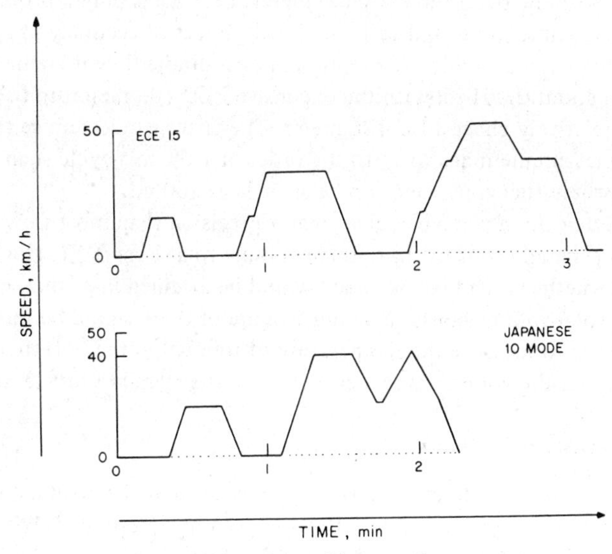

Figure 9.4 Urban driving cycles

(3) The failure to take into account the effect of transient operation on engine performance (e.g. carburetter enrichment on acceleration).

9.5.2 *Measurement of Fuel Consumed in Chassis Dynamometer Tests*

A chassis dynamometer fuel economy test has the advantage that it is possible to use a variety of methods of measuring the quantity of fuel consumed. These are the following.

(a) Measurement of fuel volume.
(b) Measurement of fuel weight.
(c) Carbon mass balance method.

9.5.2.1 *Volumetric Methods*

The measurement of the volume of fuel consumed can be made either with a simple burette or with a meter designed to measure volume flow. There are a number of commercial flow meters for measuring gasoline flow with good accuracy. These are positive displacement meters, either of the piston type (such as the Japanese Tokico meter and the US Fluidyne meter) or of the gear type (such as the German PLU meter). Accuracy of the order of 0.25–1.0% is claimed by the manufacturers, and this is generally substantiated by tests.

There are two problems in the use of flow meters in cycle tests: (i) the tendency for accuracy to be degraded at very low flow rates owing to leakage through the metering elements and (ii) the minimum increments of volume flow which can be determined by the meter. Leakage is only a problem under conditions such as idling and will only result in loss of accuracy in cycles with a high proportion of idling. The minimum measurable flow increment can vary from about 0.001 ml with the expensive PLU 103 meter up to 1.0 ml with the relatively cheap PLU 106 meter. The latter would reduce the potential best measurement accuracy to the order of 1.0% in a cycle such as the ECE 15, where fuel consumed can be as little as 100 ml.

A burette can, in principle, give greater precision than most flow meters since it is generally possible to observe volume to at least 0.2%. However, it is doubtful whether such high accuracy would be attained in normal measurements of total volume consumed, and a figure of 0.5% would be more realistic. It is essential to measure the temperature of the fuel in the burette or flow meter, so that the volume can be corrected to the standard test temperature.

9.5.2.2 *Gravimetric Methods*

The accuracy of a gravimetric method of measuring fuel consumed is limited only by the increment of fuel weight which can be determined. With conventional balances this can be as low as 0.1 g, giving a potential accuracy of

0.1% or better. Electronic weighing, involving load beams or load cells with strain gauges, are currently being used for this purpose. They have the advantage of convenience, but the resolution can be as poor[15] as 5 g, which would severely limit measurement accuracy in a cycle with low total consumption, such as the ECE 15.

A weighing method has the advantage that it can be used with cars with flow-and-return fuel systems (e.g. fuel injection cars). These present considerable problems with volume measurement because of vapour formation or degassing in the fuel return lines. With a gravimetric method there are no such problems if the flow is returned to the weighed vessel.

9.5.2.3 Carbon Balance Method

The carbon balance method for measuring fuel consumed is based on the weight of carbon contained in the exhaust constituents (principally CO_2 and CO, with HC to a lesser extent) of a known volume of exhaust gas. In addition, the weight H/C ratio of the fuel must be known. In practice, with a prescribed test fuel a fixed value can be assumed for H/C.

The weight of fuel consumed can be calculated from fuel weight,

$$g = V(1 + H/C)5.33 \times 10^{-3}\left(CO_2 + CO + \frac{HC}{10^4}\right) \quad (9.1)$$

where V is the volume of dry exhaust gas, l, at 0°C and 760 mm Hg, H/C is the weight ratio of hydrogen to carbon in fuel, CO, CO_2 are the %v concentrations in exhaust and HC is the hydrocarbon concentration, ppm C, in exhaust measured by flame ionization detection (FID) analysis. V may be derived from the measured volume V_m at t °C and B mm Hg by

$$V = V_m \times \frac{273}{(273 + t)} \times \frac{(B - P_w)}{760}$$

where P_w is the saturated vapour pressure of water at t °C, obtained from reference tables.

Equation 9.1 can be used for the calculation of the weight of fuel consumed in both the European 'big bag' and the US and Japanese constant-volume sampling (CVS) methods of exhaust analysis.

The EPA have derived a version of the equation that is specifically applicable to the calculation of fuel economy in the US emission test procedure using the UDDS[16]. The general expression obtained is

fuel economy, mile/US gal =

$$\frac{3785SG}{(H/C + 1)\left\{\left(\frac{1}{H/C + 1}HC\right) + (0.429CO) + (0.273CO_2)\right\}} \quad (9.2)$$

where SG is the fuel specific gravity, H/C is the weight ratio of hydrogen to carbon in fuel and HC, CO, CO_2 are the weight emissions, g/mile.

With the test fuel used (Indolene) H/C = 0.154 and SG = 0.739, and equation 9.2 reduces to

$$\text{fuel economy, mile/US gal} = \frac{2423}{(0.860\text{HC}) + (0.429\text{CO}) + (0.273\text{CO}_2)} \quad (9.3)$$

(1 mile/US gal = 0.425 km/l)

The carbon balance method involves certain assumptions in the derivation of fuel consumed.

(i) The only carbon in the exhaust is contained in the HC, CO and CO_2. This means that other carbon-containing compounds, such as oxygenated hydrocarbons that are not detected by an FID and carbonaceous particulates, are ignored.

(ii) There are no other sources, apart from the fuel, of HC, CO and CO_2 in the exhaust.

(iii) The carbon in the exhaust accounts for all the fuel consumed during the test, i.e. there are no vapour losses from the carburetter or accumulations of fuel or hydrocarbons in the crankcase oil, etc.

Errors arising from (i) could be significant in some circumstances; for example, when measuring economy with methanol fuel. Methanol, or its derivatives, in the exhaust will not be determined by FID analysis. It should be emphasized that hydrocarbon measurement by a hexane-sensitized non-dispersive infra-red (NIDR) analyser is basically unsuitable in the carbon balance method because it does not detect all hydrocarbons in exhaust sample. A correlation factor to relate the NDIR to the FID analysis is not fully reliable.

9.5.2.4 Comparison of Methods

It may seem that the carbon balance method is an overcomplicated technique for the measurement of fuel consumption in a chassis dynamometer test, compared with a conventional volumetric or gravimetric technique. However, it becomes more attractive when it is realized that, at present, chassis dynamometer tests for the measurement of fuel economy are generally carried out in facilities already equipped for emission measurement. This means that a very accurate CO_2 analysis is the only measurement required in addition to a conventional exhaust analysis.

The carbon balance method has the positive advantage that no modification has to be made to the fuel feed to the carburetter, and there are no problems with fuel feed systems with return flow, such as gasoline or diesel fuel injection systems.

Table 9.11 Comparison between carbon balance and volumetric and gravimetric fuel consumption measurements

Car	Method	Mean fuel economy, mile/US gal	Standard deviation	% standard deviation
2.3 l, four cylinder	Carbon balance	18.56	0.258	1.39
	Volumetric	18.62	0.480	2.58
	Carbon balance	18.36	0.480	2.61
	Gravimetric	18.24	0.646	3.54
5.75 l, V8 cylinder	Carbon balance	10.68	0.172	1.61
	Volumetric	10.96	0.183	1.67
	Carbon balance	10.64	0.372	3.49
	Gravimetric	10.72	0.333	3.11

Comparisons have been made between the carbon balance technique and conventional gravimetric methods of determining fuel economy. Results in the US[15] using the 1975 federal test procedure (the CVS-CH test) are given in table 9.11. They are based on eight tests in each comparison, and there is no evidence from these results that the carbon balance method gives a significantly different mean value or even a difference in standard deviation when compared with volume or weight measurement.

Poor correlations have been obtained in some cold-start tests, in which the carbon balance method gave lower fuel consumption. The reason for this has been traced to incomplete filling of the fuel feed system and carburetter float bowl at the start of the test, in which case additional fuel is (erroneously) measured by conventional volumetric or gravimetric methods. The effects of incomplete filling can be avoided by careful purging of the fuel system prior to test.

The data obtained from 245 tests have been compared by the EPA[16]. These showed that on average the fuel economy calculated from carbon balance was 3.3% higher than that derived by fuel weighing. A very similar result was obtained in tests at TRC with a 2 l car in the 1972 federal (CVS-C) test, where the carbon balance method gave a mean economy 3.4% higher than did fuel weighing.

The fuel which passes the piston rings and enters the crankcase oil will be responsible for a consistently higher figure for fuel consumption as measured by weighing or volume measurement than that obtained by exhaust analysis. This will occur, to a variable degree, during warm-up in a cold-start test. The losses due to fuel entering the crankcase oil will be additional to vaporization losses from the carburetter and the effect of incomplete float bowl filling.

It can be concluded that the carbon balance method is capable of giving accuracy comparable with that of conventional direct fuel weighing or volume measurement. However, there is an inherent trend for the exhaust analysis method to give a somewhat lower value for the quantity of fuel consumed.

9.5.3 The Chassis Dynamometer in Fuel Economy Measurement

The use of a chassis dynamometer in the measurement of fuel economy imposes stringent requirements on the overall performance of the system compared with that of the same equipment when it is used for emission measurement.

There are two major factors influencing measurement accuracy with a chassis dynamometer.

(a) The accuracy of following the test cycle.

(b) The matching of the dynamometer to simulate the road load and inertia of the test vehicle.

9.5.3.1 Test Cycle

The cycles are followed on the dynamometer using a 'driver's aid'. This is basically a potentiometric chart recorder with pen movement proportional to the dynamometer roll speed. The test cycle is drawn onto the chart, and the vehicle is driven so that the pen follows the trace on the moving chart. If a permanent record of the driven trace is not required, the pen can be replaced by a target. In the most recent design of 'driver's aid' (the Ono Sokki, made in Japan) the cycle is pre-drawn on a translucent chart which is tracked by a light spot. The chart speed is generally 100 or 150 mm/min.

The accuracy with which the trace is followed determines both the total distance travelled and the manner in which the car is driven i.e. the smoothness of driving of the cycle. The correct cycle distance is important, since in all current or proposed chassis dynamometer cycle tests the nominal cycle distance is used for the calculation of the fuel economy. A significant error in distance will directly influence the result, and variation in distance will reduce test repeatability.

It is impossible in practice to follow the cycle exactly, and, for example, in the US urban economy test using the UDDS a speed variation of ±2 mile/h is allowed at any point in the cycle around the specified speed. It has been calculated[15] that over the 7.5 mile CVS-C test this speed tolerance could give a distance variation of ±1 mile, a variation of 13%.

In practice it is possible to follow the cycle more accurately than this. The measurement of distance in the UDDS by dynamometer roll counting has shown that distance can vary with a standard deviation of 2%[15], in 30 tests on two dynamometers. The UDDS is a complex cycle compared with the ECE 15 cycle, and tests at TRC have shown that the mean distance in the ECE 15 cycle is as little as 0.1% different from the theoretical distance and that the standard deviation of the distance variation is less than 1%.

Table 9.12 Comparison of human and automatic drivers in chassis dynamometer fuel consumption measurement

Driver	Average human driver	Best human driver	Automatic driver
Accelerator movement, counts	278	185	150
Accelerator movement, range	30	89	24
Distance variation, % standard deviation	3.4	1.6	0.3
Fuel consumption variation, % standard deviation	–	0.93	0.47

The use of roll-count distance in place of the nominal distance did not improve the variation in fuel economy measured in the US CVS-CH test[15], and it was concluded that other factors overshadowed the effect of distance variation. However, the repeatability of measurements at TRC using the ECE 15 cycle was slightly improved when the measured distance, rather than the prescribed distance, was used in the calculation of fuel economy.

In principle, the repeatability and accuracy of fuel economy measurements should be improved if the roll-count distance is used in place of the nominal distance. Thus, to obtain the optimum measurement precision the chassis dynamometer should be instrumented to measure the roll-count distance.

The smoothness with which the driver follows the trace on the 'driver's aid' can influence fuel economy without affecting cycle distance. Jerky driving, involving unnecessary use of the accelerator, can adversely influence fuel consumption. There is a significant element of skill in cycle driving on a chassis dynamometer, and repeatability improves with the experience of the driver. Tests with a mechanized servo-feedback automatic driver system[17] have shown that the automatic driver gave fewer accelerator movements and improved repeatability of distance and fuel consumption measure measurement, as shown in table 9.12.

The application of automatic driver systems is practicable with cars fitted with automatic transmissions and automatic chokes, as is general in the US. In the European situation, with most cars having manual transmissions and many with manual chokes, the greatly increased complexity of an automatic system renders this route to improved precision less attractive.

9.5.3.2 Simulation of Road Conditions

The chassis dynamometer must be capable of reproducing the various forces acting on the car when driven on the road. This means that the dynamometer must reproduce the road-load forces, rolling resistance and aerodynamic drag, and the inertia forces.

Inertia On most dynamometers, vehicle inertia is simulated by sets of flywheels, so that the inertia weight can be varied to match the car. At present, the majority of chassis dynamometers used for fuel economy measurement were developed for emission measurement and have minimum inertia weight intervals of 100 kg, and in some cases 200 kg. It is considered that these intervals are too large even for heavy US cars, and this is even more so when considering the relatively light-weight European or Japanese car. For economy measurement the minimum inertia intervals should be not more than 50 kg.

Road-load forces While it is relatively simple to simulate inertia forces on a chassis dynamometer, the simulation of the road-load forces is by no means as simple. Again, part of the difficulty arises because the dynamometers were originally used for emission measurements. The accuracy of road-load matching is not particularly important in an emission test, and many emission dynamometers use relatively simple water brakes. In fact in the US federal emission test it is permitted to use a dynamometer load setting corresponding to the vehicle weight, the so-called 'cook book' method established in the *Federal Register*. While this has proved satisfactory for emission testing, it is not acceptable in a fuel economy test, since it does not take into account differences in rolling resistance or aerodynamic drag for vehicles of the same weight. As an illustration, it is only necessary to consider the case of a vehicle such as the Land-Rover in comparison with a low-drag coupe to appreciate that a common road load would introduce substantial errors.

In fuel economy measurement the dynamometer must match the load–speed characteristics of the car on the road. The load–speed characteristic has to be determined on the road, and the brake then has to be set to reproduce the required load pattern. In reality this cannot be fully achieved with a simple water brake, and the practice is to match the load at a single speed appropriate to the cycle, e.g. 50 mile/h in the UDDS and 50 km/h in the ECE 15 cycle. The errors which may exist at lower speeds are ignored, which is tolerable for emission measurements. In urban cycles, with relatively little cruising and considerable acceleration, most of the work is done against inertia forces[18], and errors in the simulation of road load may not significantly effect the result of economy measurements. Unfortunately, this is not true for a highway cycle; for example, in the EPA highway cycle, 65–75% of the energy is used against the road-load forces, rolling resistance and aerodynamic drag. The correct measurement of fuel economy in highway cycles requires the accurate simulation of road load over a relatively wide speed range, e.g. 30–60 mile/h in the EPA highway cycle.

For the accurate measurement of fuel economy the chassis dynamometer should be equipped with an electric brake, either a d.c. or an eddy current machine, for in principle an electric brake can be programmed to match any required load–speed characteristic. Unfortunately an electric brake is usually much more expensive than a water brake.

The performance of the brake in the simulation of road-load forces will be no better than the accuracy of the load data obtained from the road. In the US for more exact work than federal emission tests, it is general practice to match the brake so as to give the same coast-down or run-down times as those obtained on the road. This is essentially an averaging technique, in that road load is matched over a speed range, e.g. 10 or 20 mile/h, and the road load is not necessarily matched at a particular speed. The method has the advantage that no instrumentation is required to obtain the road data, apart from a stop-watch. It has the disadvantage that the correct setting of the dynamometer can only be achieved after a number of tests at different brake loads and is thus time consuming.

In the proposed European urban fuel economy test[7] the manifold vacuum method is used to obtain the road-load matching. In this technique manifold vacuum is measured at 50 km/h on a level road in two-way runs, and the brake load is adjusted to give the same vacuum at 50 km/h on the dynamometer. Experience has shown that the accurate measurement of manifold vacuum on the road is by no means easy, and recording of the vacuum over a period at steady speed would appear to be desirable.

The correct matching of the dynamometer is of extreme importance in the measurement of steady-speed fuel consumption on a chassis dynamometer. For example, the European proposals for the measurement of fuel economy[7] specify measurement at 90 and 120 km/h either on the road or on a chassis dynamometer. To obtain equivalent values it will be essential to match the dynamometer accurately at these speeds. The conventional coast-down method is inherently unsuitable for this purpose, and careful application of the manifold vacuum technique will be required.

The load matching of diesel-engined passenger cars poses special problems. The manifold vacuum method is not applicable, for manifold pressure does not change with load. Preliminary tests at TRC have shown that a diesel car can be load matched by measuring exhaust CO_2 on the road and by adjusting the dynamometer brake to give the same value. (The exhaust CO_2 increases with road speed.) The coast-down technique can also be used.

The correct road load for a vehicle can be calculated from published data on the rolling resistance, frontal area and drag coefficient. While this method is sound in principle, in practice it is vitiated by the lack of sufficient vehicle data, doubt about the precision of the data, the possibility that data obtained on one car will not apply to all examples of that car and uncertainties about vehicle behaviour on the dynamometer.

There are two further problems relating to the use of chassis dynamometers for fuel economy measurement. These are the roll diameter and the cooling-air system. Many chassis dynamometers developed for emission measurements use small-diameter twin-roll systems and relatively small constant-speed

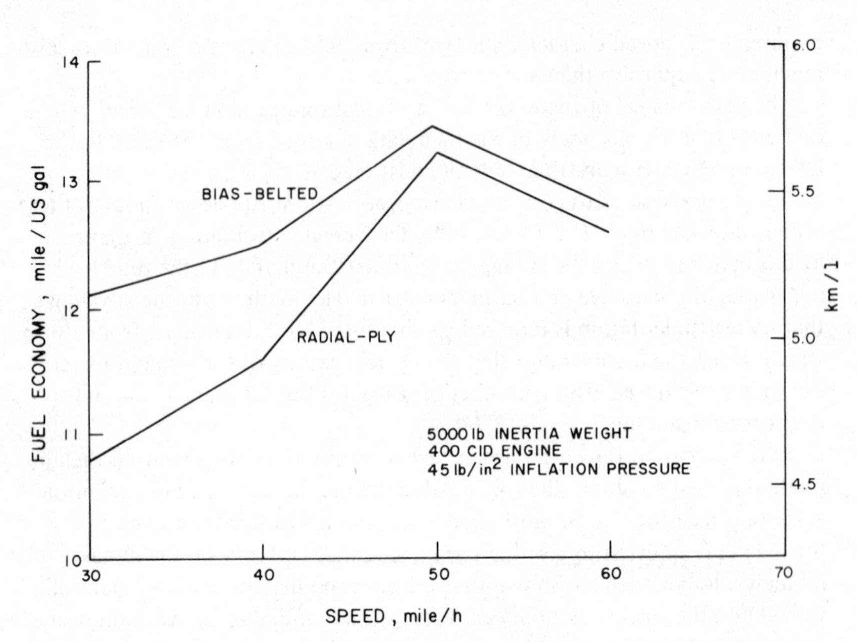

Figure 9.5 Steady-state fuel economy versus dynamometer speed for bias-belted and radial-ply tyres[15]

electric cooling fans. The Clayton dynamometer normally used in the US has twin rolls (22 cm in diameter) with centre spacings of 50 cm. Fuel consumption measurements with a Clayton unit[15] show that fuel economy is poorer with radial-ply tyres than with bias-belted or cross-ply tyres, as is shown in figure 9.5. This is the reverse of the road situation, and measurements on the dynamometer show that there is greater internal power dissipation on the dynamometer rolls with radial-ply tyres.

The effect of roll diameter and roll spacing is of increased significance with the small wheels and tyres common on European and Japanese cars. It is desirable that the behaviour of the tyres on the chassis dynamometer should be the same as that on the road. This is most nearly achieved with large single-roll dynamometers, but this type is not as convenient to use as twin-roll units and has considerably higher capital and installation costs. A twin-roll unit with closely spaced rolls of not less than 50 cm in diameter is a reasonable compromise, and this type is recommended as an alternative to large single-roll machines.

The use of a fixed-speed fan system with open bonnet, the normal practice with most emission dynamometers, gives a different pattern of air flow, particularly around the engine, than that obtained on the road. This means that effects of vehicle warm-up may differ from those on the road. It is desirable that the cooling fan should give air flow proportional to vehicle speed and that the fan outlet area should approximate to the frontal area of the bonnet. With this type of cooling-air system the vehicle is driven with a closed bonnet, and warm-up and under-bonnet temperatures are a better match to road conditions. Unfortunately, a system of this type is much more expensive and bulky than the more common fixed-speed fan.

9.5.4 The Variation of Cycle Test Fuel Economy Measured on Chassis Dynamometers, and a Comparison with Road Measurements

A considerable amount of statistical data has been published on measurement variation in the US urban economy test cycle. There is comparatively little information presently available on measurement in the ECE 15 cycle.

The EPA has reported[19] that the standard deviation of 30 measurements in the same test cell in the CVS-H test with a Ford 6.91 car, using the carbon balance method, was 2.6%. The fuel economy of a 1973 Chrysler measured by the CVS-H procedure using the carbon balance method in five different test cells[19] gave the results shown in table 9.13.

The results obtained in correlation tests with the same vehicle at a number of laboratories[19] are given in table 9.14. This shows results obtained by the CVS-H test using the carbon balance method.

The results obtained in chassis dynamometer tests using the UDDS in the CVS-H procedure show that the standard deviation for measurements at individual laboratories can vary from 1% at best to 5% at worst. It is difficult to establish a mean value generally applicable to measurements at a single laboratory, but it would appear that a standard deviation of 1.5–2% represents a realistic figure for repeatability. Table 9.14 would indicate that the reproducibility, i.e. the variation for measurements at different laboratories, has a standard deviation of about 3%.

The comparative figure for an urban cycle test with cars driven on the road (see table 9.5) indicates that the chassis dynamometer test gives larger variations. However, it should be remembered that the UDDS chassis dynamometer cycle is very much more complex than the SAE urban road test cycle.

There are very few comparable data for the European urban economy test using the ECE 15 cycle, but tests have been carried out at TRC to compare the fully warmed-up fuel economy in this cycle on an emission-type chassis

Table 9.13 Variation of fuel economy in the CVS-H test with the same car in five test cells

No. of tests at each cell	Mean fuel economy for each cell, mile/US gal	Range of % standard deviation at each cell
13–18	10.83–11.51	2.8–5.5

Table 9.14 Fuel economy in the CVS-H test measured with three cars at different laboratories

Model year, engine type	1973, 2.6 l V6		1974, 5.7 l V8		1974, 7.5 l V8	
Laboratory	Mean, mile/US gal	% standard deviation	Mean, mile/US gal	% standard deviation	Mean, mile/US gal	% standard deviation
EPA	20.0	2.7	12.6	1.9	10.5	5.8
Ford laboratory 1						
Test 1	19.7	2.8	12.5	1.0	10.5	1.5
Test 2	19.7	3.0	12.4	1.2	10.6	2.2
Ford laboratory 2	19.6	1.5	12.6	1.5	10.7	0.9
Ford laboratory 3	18.7	1.1	–	–	9.7	1.6
California Air Resources Board	18.9	1.2	–	–	10.5	1.6
Overall	19.5	3.3	12.5	1.6	10.4	4.9

Table 9.15 Repeatability of measured fuel economy in consecutive, single, hot ECE 15 cycles on the chassis dynamometer and on the test track

Measured fuel economy, l/100 km	Chassis dynamometer test	Road track test
Maximum	11.65	11.98
Minimum	13.39	12.74
Mean	12.13	12.26
% standard deviation	5.0	2.1

dynamometer and on a road test track. A 1.6 l car was used, and consumption was determined gravimetrically on the dynamometer and with a flow meter on the test track. The cycle was followed on the track using a tape recorder as a 'driver's aid'. The results are shown in table 9.15.

Table 9.15 shows that the variation in the test track measurement was less than half that on the dynamometer. This may appear to be a puzzling result, but it is in line with indications from US work, and it reflects the repeatability of the work done by the car in the two cases. In the road test the forces acting on the car were the same for each cycle (the wind velocity was very low), whereas in the dynamometer test the forces would only be the same if the operation of the water brake was completely repeatable. The fact that the mean consumption differed by only about 1.0% in the two tests would indicate that the problem is one of brake stability rather than a consistent difference in vehicle loading.

These results show that the variation in the proposed European urban economy test, in which consumption is measured over a total of six ECE 15

Table 9.16 Fuel economy of six cars measured in the EPA highway cycle on a chassis dynamometer and on a road test track

Car capacity, inertia weight	Fuel economy, mile/US gal	
	Dynamometer test	Road track test
3.7 l, 3500 lb	25.7	22.6
5.2 l, 4000 lb	20.9	17.5
5.9 l, 5000 lb	18.1	16.6
5.9 l, 5000 lb	18.6	15.9
6.7 l, 5000 lb	17.4	16.2
6.7 l, 5000 lb	14.6	14.0
7.2 l, 5000 lb	17.6	16.4

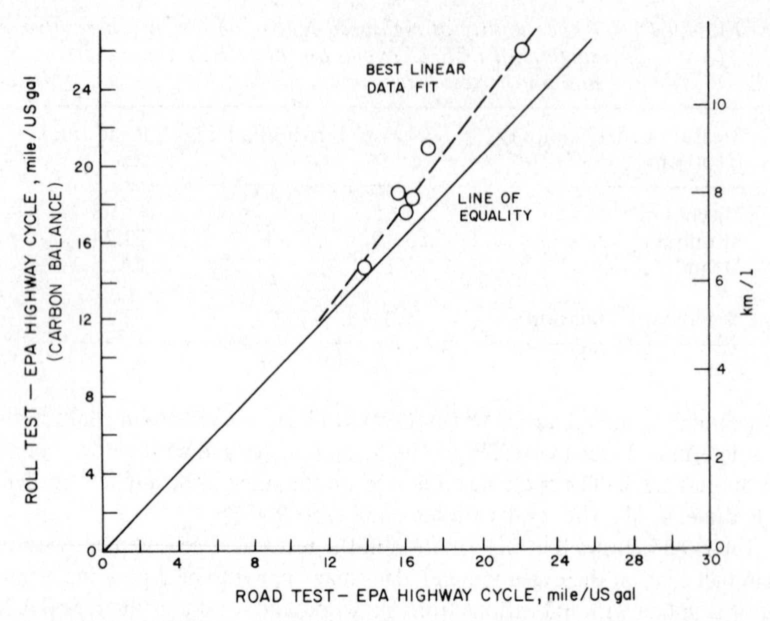

Figure 9.6 Fuel economy for EPA highway cycle as measured on a track and a dynamometer[19]

cycles, would have a standard deviation of 2.0% and 0.9% for the chassis dynamometer and road track tests, respectively.

It has been pointed out earlier that the matching of the road load on a dynamometer is more important in a highway test than in an urban cycle test. This view is supported by a comparison of chassis dynamometer and road track measurements with the EPA highway cycle[19]. The results were obtained on a Clayton dynamometer with road load set by the *Federal Register* 'cook book' method and are shown in table 9.16 and figure 9.6.

The results show that better economy is obtained, in all cases, with the dynamometer; the differences range from 4% to 16%. Figure 9.6 indicates that the difference increases as fuel economy increases, and this may reflect the poorer matching of the dynamometer as vehicle weight (and size) decreases.

The results discussed in this section highlight the critical role of brake performance in cycle fuel economy tests using chassis dynamometers.

9.5.5 Comparative Advantages and Disadvantages of Chassis Dynamometer and Road Track Tests

We are now in a position to identify the relative advantages and disadvantages of chassis dynamometer and road track tests for the measurement of fuel economy.

The advantages of road or tack testing are as follows.

(1) The total load, including inertia, and road loads are accurately simulated.

(2) Fuel economy tests may be carried out on any suitable level road.

(3) Relatively low expenditure on equipment is required, providing an appropriate road or track is available.

The disadvantages of road or track testing include the following.

(1) Tests cannot be carried out under adverse weather conditions.

(2) The measured fuel economy has to be corrected for variations in ambient conditions.

(3) It is difficult to drive complex cycles on the road in a consistent manner.

(4) Cold-start tests are unsatisfactory if ambient temperatures vary widely.

(5) Fuel economy and exhaust emissions cannot be measured simultaneously.

(6) Fuel consumed has to be measured by volumetric techniques (although in some cases gravimetric methods can be used).

(7) A suitable test track may not be available locally.

The advantages of a chassis dynamometer test include the following

(1) The test can be carried out independently of local weather conditions.

(2) Correction factors for the effect of ambient conditions are minimized, since test conditions can be controlled.

(3) Measurements can be made over a wide temperature range (in a temperature-controlled chassis dynamometer).

(4) There is no limitation on the complexity of the test cycle, since the driving cycle can be followed within closely specified limits.

(5) Fuel economy and exhaust emissions can be measured simultaneously.

(6) All techniques for the measurement of fuel consumed can be used.

The disadvantages of a chassis dynamometer test include the following.

(1) The total load simulated on the chassis dynamometer may not duplicate the rolling resistance and aerodynamic drag of the vehicle on the road.

(2) The air flow of the cooling fan may not reproduce the air flow pattern of the moving vehicle. Effects of vehicle warm-up and under-bonnet temperature may differ from those on the road.

(3) Large intervals in inertia weight on some chassis dynamometers may give rise to incorrect inertia loading of the vehicle.

On balance the advantage would appear to lie with the chassis dynamometer test. This is particularly so when investigating the effects of fuels or lubricants on fuel economy. In these cases the ability to control the test temperature is of paramount importance. In many cases fuel or lubricant studies must be carried out at low temperatures, when a fully temperature-controlled dynamometer is mandatory. The divorcing of economy testing from local weather conditions is an additional powerful advantage of the chassis dynamometer.

The major problem with the chassis dynamometer test at present is the poor performance of the dynamometer itself. Both the accuracy and stability of simulation of vehicle load are unsatisfactory on dynamometers commonly used for emission testing. Only an electric brake of advanced design is suitable for this application.

9.6 Measurement of Fuel Economy in Bench Engine Tests

The measurement of brake specific fuel consumption (bsfc) in a bench engine test is fundamental as an assessment of engine thermal efficiency, but, at the same time, it is the measurement least relevant to road fuel consumption.

In a test of this type the operating conditions of the engine are well defined and controlled, but it is difficult to relate these conditions to the operation of a car on the road. For example, it is necessary to know the power absorption of the engine auxiliaries and the transmission components, gearbox, final drive and tyres.

The measurement of bsfc involves the simultaneous measurement of the rate of fuel input to the engine, the brake load and the engine speed. To obtain an accurate figure, the quantity of fuel consumed should be measured and the engine revolutions counted over a suitable time period, thus giving an averaged engine speed. It is desirable to chart-record the load on the engine brake with great accuracy over the test period to establish a mean load.

If an electric brake capable of motoring the engine is used, the indicated specific fuel consumption (isfc) can be determined. This technique may not give precisely the same value of isfc as that calculated from $p-V$ diagram, the classical method, since engine friction is influenced by the cylinder pressure, i.e. it is increased when firing, compared with motoring.

Paradoxically, there are problems in the definition of bsfc which do not arise when considering vehicle consumption. The value obtained for bsfc will depend on the engine auxiliaries fitted to the engine in the bench test, since these absorb significant amounts of power. The engine fan, generator, etc., are not essential in a fully equipped bench test installation. The extent to which the auxiliaries are included in the measurement of bsfc is defined in the various engine test specifications, e.g. the SAE gross and net measurements[20] and the Deutsches Institut für Normung procedure[21].

Both gravimetric and volumetric methods can be used to determine the quantity of fuel consumed in a bench test; fuel weighing is the most satisfactory since it is both rather more accurate and does not require temperature correction.

Table 9.17 A comparison of fuel economy test methods

Test type	Practical relevance	Precision		Fuel flow measurement method[b]	Practicability	
		Repeatability	Reproducibility		Equipment cost	Total cost
Uncontrolled road tests	Very high	Very poor	Very poor	V,GS	Very low	Very high[c]
Controlled road tests	High	Fair	Poor	V,GS	Low	Very high[c]
Road cycle tests	High[a]	Good	Fair	V,GS	Low[d]	Medium[d]
Chassis dynamometer cycle tests	High[a]	Good	Fair	V,GC,CB	High	High
Bench engine tests	Low	Good	Fair	V,GC,CB	High	High

[a]Provided representative driving cycles are chosen.
[b]Fuel flow notation: V, volumetric; GS, gravimetric (single weighings before and after the test); GC, gravimetric (single weighings and continuous measurement throughout the test); CB, carbon balance by exhaust analysis.
[c]For a given level of precision.
[d]Discounting very high cost of test track.

9.7 Concluding Remarks

The experimenter setting out to measure fuel economy is currently faced with a very wide choice of methods, made wider now than a decade ago by the advent of test cycles and chassis dynamometer procedures. Table 9.17 summarizes the comparative merits of the five alternative types of test which have been discussed in this chapter.

It is evident from this table that it is not easy to choose one method with unique advantages. For instance, there is clearly a conflict between the relevance of a test to the practical motoring situation and the ease of control and cost of the test.

There is a clear trend towards the adoption of chassis dynamometer tests for the measurement of fuel economy, their potential advantages being good precision and control, which can only be realized by the development and adoption of purpose-built high-quality chassis dynamometers and the use of representative cycles. On balance this method appears to hold the most promise for the future.

References

1. Fakes and claims. *Motor* (11 October 1975) 43
2. R. F. Barker. The trip-mileage analyser. Unpublished Shell work (1974)
3. The development of the new SAE motor vehicle fuel economy measurement procedures. *Soc. Automot. Eng. Pap.*, No. 750006 (1975)
4. Automobile fuel economy: voluntary labelling programme. *Fed. Regist.*, **38** (27 August 1973) part II, 165
5. R. E. Kruse and T. A. Huls. Development of the federal urban driving cycle. *Soc. Automot. Eng. Pap.*, No. 730553 (1973)
6. Measurement of the conventional fuel consumption of passenger cars. *French Govt. Stand.*, No. NFR 11-502 (April 1975)
7. Draft recommendation for the measurement of fuel consumption of vehicles. Proposal by the Groupement des Rapporteurs de la Pollution de l'Air to the Inland Transport Committee of the Economic Commission for Europe (1975)
8. Fuel economy measurement – road test procedure. *Soc. Automot. Eng. J.*, No. 1082 (April 1974)
9. M. Smith and M. J. Manos. Determination and evaluation of urban vehicle operating patterns. *Air Pollution Control Assoc. Pap.*, No. 72-177 (1972)
10. *Fed. Regist.*, **38**, No. 84 (2 May 1973)
11. New motor vehicles and engines; air pollution control – 1974 model year test results. *Fed. Regist.*, **39**, No. 40 (27 February 1974) 7664
12. T. C. Austin, R. B. Michael and G. R. Service. Passenger-car economy trends through 1976. *Soc. Automot. Eng. Pap.*, No. 750957 (1975)

13. R. E. Kruse and C. D. Paulsell. Development of highway driving cycles for fuel economy measurements. *US Environ. Prot. Agency, Emission Control Technol. Div., Ann Arbor, Mich., Rep.* (March 1974)
14. C. D. Paulsell. Amendments to the report on development of a highway driving cycle for fuel economy measurements. *US Environ. Prot. Agency, Ann Arbor, Mich., Rep.* (April 1974)
15. B. H. Simpson. Improving the measurement of chassis dynamometer fuel economy. *Soc. Automot. Eng. Pap.*, No. 750002 (1975)
16. T. C. Austin and K. H. Hellman. Passenger-car fuel economy – trends and influencing factors. *Soc. Automot. Eng. Pap.*, No. 730790 (1973)
17. R. T. Gryce. Ford auto/emission driver system. *Soc. Automot. Eng. Pap.*, No. 741007 (1974)
18. C. LaPointe. Factors affecting vehicle fuel economy. *Soc. Automot. Eng. Pap.*, No. 730518 (1973)
19. Emissions and fuel economy test methods and procedures. *US Environ. Prot. Agency, Wash., Consult. Rep.*, presented at *Comm. on Motor Vehicle Emissions Commission on Sociotech. Syst., September 1974*
20. Engine test code – spark ignition and diesel. *Soc. Automot. Eng. J.*, No. 816B (1975) part II
21. Vehicle construction – engine output net and gross. *Dtsch. Inst. Normung Procedure*, No. V70020 (December 1973)

10 The Effect of Crankcase Lubricants on Fuel Economy

B. BULL and A. J. HUMPHRYS

10.1 Introduction

The internal combustion engine, whilst more efficient than its predecessor the steam engine, is still only capable at present of converting about a quarter of the energy available from its fuel into useful work. The remaining three-quarters are lost as heat in the exhaust gases, in the cooling water and in removal from the external surfaces of the engine (see also appendix E).

Frictional forces inherent in the operation of the gasoline engine account for a fair proportion of the energy that is not converted into useful work. Early investigations[1] in this field established that in an engine operating under full-load conditions the losses due to friction amount to about 10% of the total energy input. Furthermore, as the magnitude of the frictional losses is relatively independent of engine operating conditions, their significance increases as the power output of the engine decreases. Under low-load conditions, losses due to friction can account for up to 40% of the energy input to the engine. Other investigators have stated[2] that an energy saving of at least 25% would be possible if all sources of friction in an engine could be eliminated. Whilst the prospect of total elimination of engine friction is a somewhat optimistic dream, there nevertheless appears to be scope for significant improvement of engine efficiency by the reduction of friction.

It will be appreciated that the bulk of the energy committed to overcoming internal engine friction can be attributed to the shearing of oil films between working surfaces. The major contributions are thought to originate where shearing stresses are highest, such as at the interface between the piston or piston rings and cylinder walls, and at the crankshaft journals. Other losses are incurred as a result of pumping and churning of the oil. Figure 10.1, derived from reference 3, shows the results of an analysis of the power losses at wide-open throttle (WOT) in a 1.5 l gasoline engine, which were obtained by motoring the engine in various stages of disassembly. As the authors point out, the absolute accuracy of such measurements is questionable because the temperatures and pressures in the cylinders of a motored engine are quite different from those when the engine is firing. Nevertheless, the relatively large contribution of 'friction' losses, which is greater than 50% of the total losses, is unlikely to be greatly in error. The effective viscosity of the lubricant in these various situations thus could be expected to have a significant influence on frictional energy losses, and it follows that reduction of this effective viscosity should lead to energy savings. However, the scope for improving engine efficiency by this means is limited, for excessively low lubricant viscosity can exacerbate lubrication problems such as oil consumption, engine wear and piston cleanliness.

For marketing purposes, the viscosity of crankcase oils is generally expressed in terms of the viscosity classification of the Society of Automotive Engineers (SAE). In this system, oils are graded into eight categories according to their viscometric characteristics. Four of the categories (5W, 10W, 15W and 20W) are defined by viscosity measurements at 0 °F (255 K). Oils in these categories are intended for winter use. The remaining four categories (20, 30, 40 and 50) are defined by viscosity measurements made at 210 °F (372 K). In recent years, oils have been developed which simultaneously satisfy the requirements of two categories. These are designated by the appropriate two

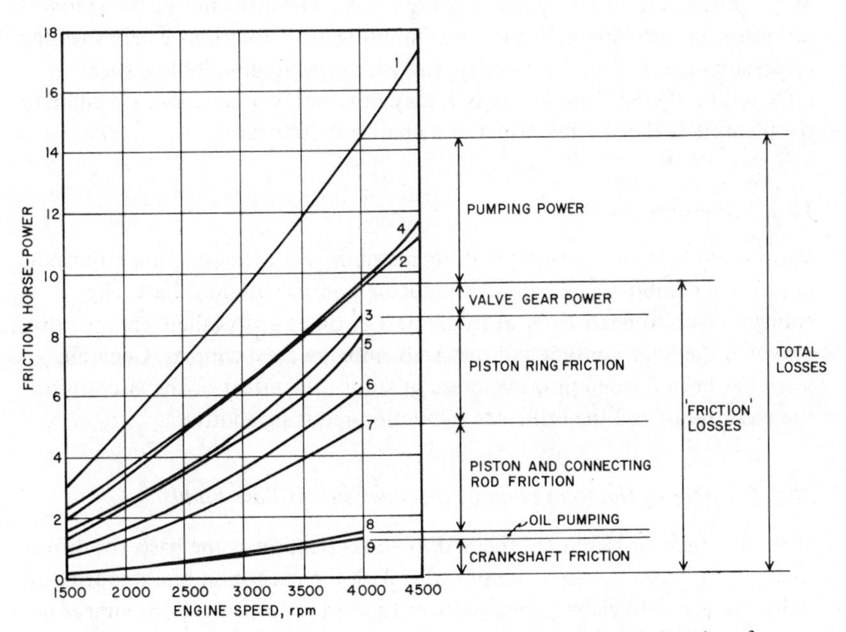

Figure 10.1 Analysis of engine power loss[3] for 1.5 l engine with oil viscosity of SAE 30 and jacket water temperature of 80 °C (353 K): curve 1, complete engine; curve 2, complete engine with push rods removed; curve 3, cylinder head raised with push rods removed; curve 4, as for curve 3 but with push rods in operation; curve 5, as for curve 3 but with top piston rings also removed; curve 6, as for curve 5 but with second piston rings also removed; curve 7, as for curve 6 but with oil control ring also removed; curve 8, engine as for curve 3 but with all pistons and connecting rods removed; curve 9, crankshaft only (1 hp = 0.746 kW)

category numbers, e.g. SAE 10W/30, 20W/40, and are termed multi-grades. Pure mineral oils are not normally able to fulfil the multi-grade requirements and thickening agents are added to them in order to achieve the desired viscometric characteristics. These thickening agents (known as viscosity index (VI) improvers) are polymeric materials which have little or no thickening effect at low temperatures. At high temperatures the effect increases markedly, permitting the viscometric requirements to be met. A multi-grade oil is therefore blended from base stocks with much lower viscosities at low temperatures than those of a single-grade oil; and, since fuel economy is so adversely affected by the cold start, a significant gain in fuel economy is expected for a multi-grade oil compared with its analogous single-grade oil (e.g. 10W/30 compared with 30 grade).

A major shortcoming of the SAE classification is that the viscometric characteristics of the oil under high shear are not considered. This is a particularly important omission in the case of a multi-grade oil that contains polymeric VI improver. When such an oil is subjected to high shearing stresses, its effective viscosity will be lower than would be expected from its SAE grade number. This is a result of the temporary reduction of the thickening power of the polymer induced by the shearing forces. The influence of the crankcase lubricant on fuel economy is greatest in lubricating situations where shearing forces are high. It thus follows that viscosity measured under low shear, as reflected by the SAE grade number, may not always be an adequate guide to the possible fuel economy effect of a crankcase lubricant.

10.2 Literature Survey

Various workers have investigated the possibility of increasing the efficiency of internal combustion engines by reducing internal friction. Naturally enough, effort appears to be at its greatest at times of so-called 'energy crisis', of which the Suez conflict and the Arab embargo are examples. Generally, work has been divided into two areas of study: the effect of the viscosity of the base fluids, and the influence of friction-reducing additives.

10.2.1 Effect of the Reduction of the Viscosity of Base Fluids

Most investigators have used the SAE oil classification as the basis for differentiating between the viscosities of oils. With the exception of a minority of rather exaggerated claims, the results of the various investigations appear to be in general agreement.

Whitehouse and Metcalfe[4] carried out experiments on both gasoline and diesel engines on test beds. They compared SAE 5W with SAE 30 oils and observed fuel consumption benefits when using the less viscous 5W oil of up to 10% under low-load conditions decreasing to 4–5% at full load.

In 1955, the Thornton Research Centre (TRC) conducted a series of road tests to assess the possible fuel consumption benefits resulting from the introduction of SAE 10W/30 oils. Compared with an SAE 30 grade oil, it was shown that average savings were 15% during slow-speed running and 3½% during prolonged high-speed running.

Also in 1955, BP launched an 'all-season' multi-grade oil[5]. They claimed that their oil, an SAE 10W/30 grade, could give substantial fuel economy benefits when used in place of SAE 30 grade lubricants. Average gains quoted were 7% for short-trip driving and 4% for high-speed tests.

Recent investigations by Lubrizol in the US[6] have emphasized the importance of the influence of trip length on fuel economy. Figure 10.2 shows[5] the improvement in fuel economy given by a 'thin' SAE 10W/40 oil (oil C) over a 'thick' 10W/40 oil (oil A) in cold-start trips of lengths of up to 8 miles (13 km). The thinner oil gave a 3.2% improvement in a 1 mile (1.6 km) trip and 1.9% improvement in a trip of 8 miles (13 km). These studies confirm

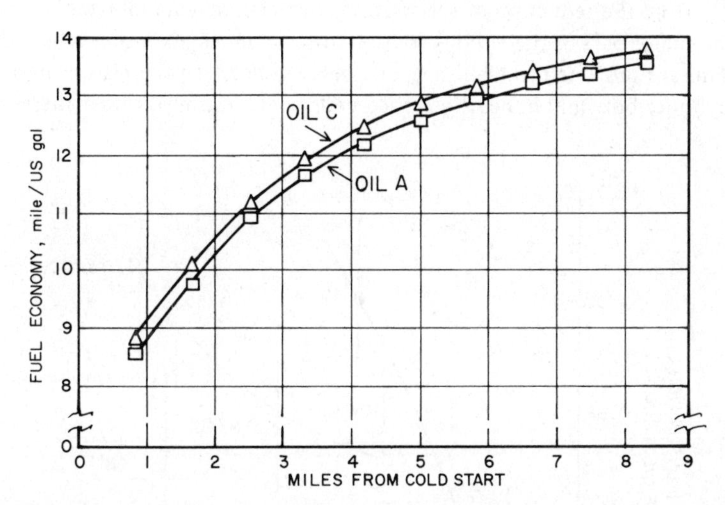

Figure 10.2 Engine oil comparison on cold-start fuel economy of compact sedans[6] (140 cubic inch displacement (CID) four-cylinder engines). Results corrected to ambient temperature of 35 °F (275 K)

Viscometric properties of test oils

		Oil A	Oil C
Viscosity grade		10W/40	10W/40
Viscosity at 210 °F,	cSt	14.59	14.94
Viscosity at 100 °F,	cSt	101.75	86.80
Viscosity at 0 °F,	P	24	12.2
VI		159	193

It should be noted that the same performance additive treatment was used in each oil. (1 mile/US gal = 0.425 km/l; 1 mile = 1.609 km; 1 in^3 = 0.0164 l)

that the benefits that result from the use of low-viscosity oils are greatest on short trips, especially when cold starting is involved. Other work by Lubrizol[7] has shown that replacing an SAE 40 grade crankcase oil by an SAE 10W/40 could result in fuel economy benefits of over 2% under warmed-up steady-driving conditions. This latter work was concerned mainly with comparisons of combinations of crankcase, transmission and axle lubricants. The contribution of the crankcase lubricant itself to the fuel economy benefits observed was not specifically studied.

A similar integrated investigation in US cars was undertaken by Davison and Haviland[8]. They concluded that reducing the viscosity of engine oil, rear-axle lubricant and automatic transmission fluid could improve fuel economy by 2–5% depending upon the type of driving. The nature of the experiments carried out made the isolation of individual fuel economy effects rather difficult. However, a small number of supplementary test data led the authors to conclude that the fuel economy benefits they observed resulted mainly from the reduction of the viscosity of the crankcase lubricant.

The major part of the viscous friction losses in an engine occur where the oil films are subjected to high shear stresses, e.g. between the piston rings and the cylinder bore and in heavily loaded crankshaft bearings. Under these

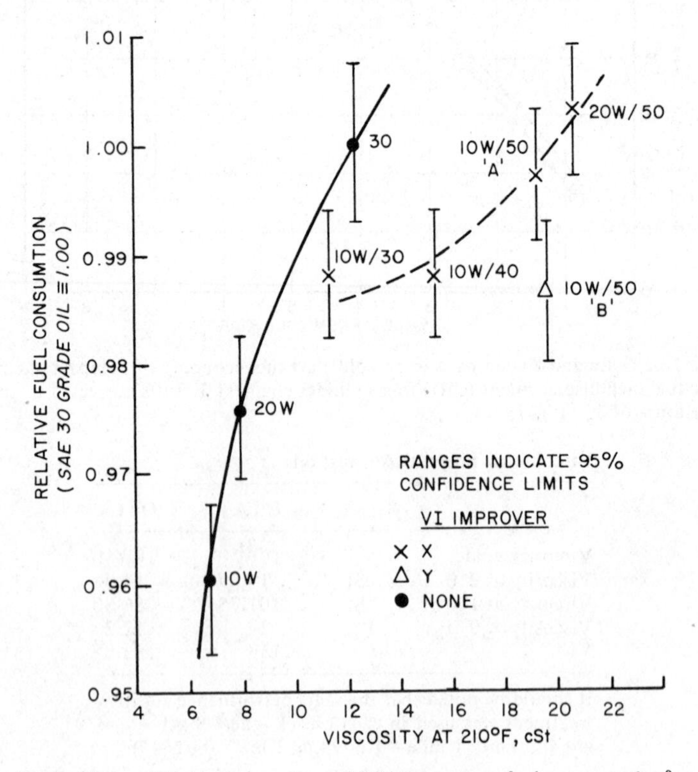

Figure 10.3 Effect of viscosity and VI improver on fuel consumption[9]

conditions polymer-thickened oils undergo a loss of viscosity which is in part reversible. The reversible loss is known as temporary shear loss. It is caused by the polymer molecules orienting themselves in the direction of the shearing force, thus reducing their resistance to oil flow and causing a local loss of effective viscosity. Temporary shear loss must be distinguished from the irreversible permanent shear loss which occurs when the molecules are ruptured by the shear stress. Modern VI improvers are designed to give minimal permanent shear loss in order to prevent loss of viscosity in engine service. However, different types of VI improver can exhibit different temporary shear characteristics, and this might be expected to affect fuel economy. Recent work at TRC using a test bed engine run at constant speed and load conditions has indicated that these temporary shear effects are significant[9]. In these tests the fuel consumptions with multi-grade oils, formulated with two different types of VI improver, were compared with those of single-grade oils. Figure 10.3 shows the relationship between fuel consumption (relative to an SAE 30 grade oil) and viscosity measured at low shear rate, as used to assess SAE grade. Figure 10.4 gives the relationship with estimated viscosity

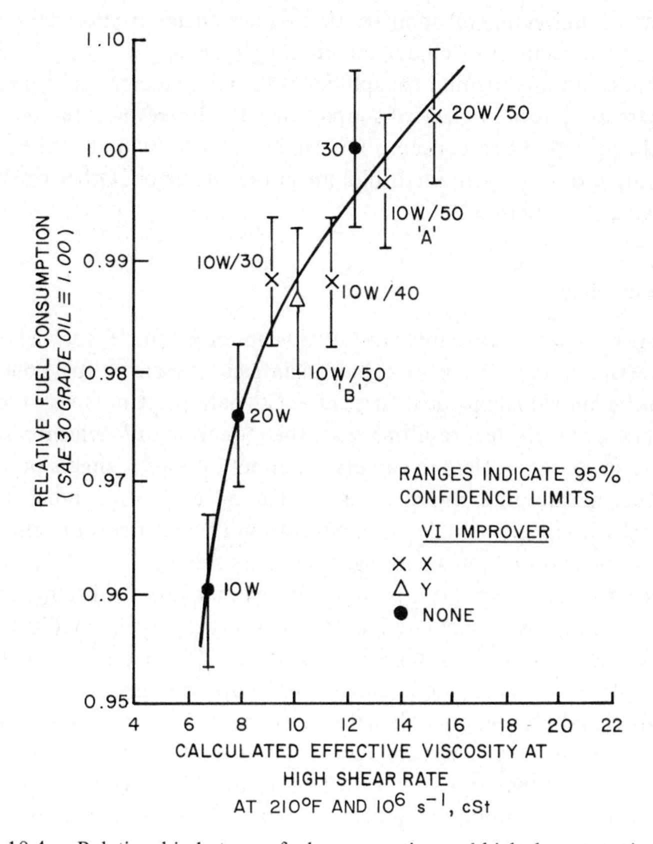

Figure 10.4 Relationship between fuel consumption and high-shear-rate viscosity[9]

at a shear rate of 10^{-6} /s which was considered to be representative of the rates prevailing in heavily loaded lubricated contacts. At this shear rate both types of VI improvers suffer significant temporary shear losses. The single relationship obtained between fuel consumption and high shear rate viscosity supports the concept that the energy losses occur mainly in high-shear-stress regions of the engine and that temporary shear effects are significant. It should be noted that the two SAE 10W/50 oils A and B gave fuel consumptions intermediate between those of the SAE 20W and SAE 30 single-grade oils and also that oil B which has a lower effective viscosity than oil A gives a lower fuel consumption.

Although the above work has indicated that the use of VI improvers which suffer large temporary shear losses can confer fuel economy benefits, there is evidence that the temporary loss of viscosity is also reflected in higher engine wear rates. It will be necessary to take this into account when evaluating the total operating economics of using less-shear-stable multi-grade oils.

10.2.2 The Effect of the Incorporation of Friction-reducing Additives

The study of lubricating oil additives that reduce engine friction does not appear to have been very well documented. Early, exaggerated claims for the effectiveness of molybdenum disulphide (MoS_2) in reducing fuel consumption were discredited owing to lack of supporting data. However, it has recently been claimed[10] that fuel consumption can be reduced by up to 4% by the incorporation of molybdenum disulphide in the engine oil. Other workers[11] have cited a 2% benefit.

10.3 Discussion

Many other variables associated with driving a motor vehicle, such as traffic density, weather and driving style, have significant effects on fuel consumption. To the individual motorist these effects would probably mask the beneficial fuel economy effect resulting from the reduction of lubricant viscosity. However, on a larger scale, the cumulative effect of large numbers of incremental fuel consumption improvements could indeed be discernible. This could be the case, for example, in the operation of large fleets of vehicles or when fuel consumption on a national scale is considered.

In the national context, it is particularly relevant that a substantial proportion of mileage is accumulated under cold-start short-trip conditions. Surveys carried out[12,13] have indicated that 75–80% of US trips are less than 10 miles (16 km) in length, and it is on short-trip operation that lubricant-related fuel economy benefits are expected to be greatest. Such short trips are said to account for about 50% of the US national gasoline consumption, and there thus seems to be considerable scope for improving the 'national fuel economy' by judicious choice of engine lubricants. A cautionary corollary must, however, be added to this statement. Important lubricant performance characteristics such as wear protection and oil consumption are related to the viscosity of the lubricating fluid. A lubricant which has an effective viscosity below certain limits will undoubtedly give rise to excessive engine wear and probably cause increased oil consumption. The penalties resulting from the use of such an oil could conceivably outweigh any associated fuel economy benefits. It should also be noted that the possible side-effects of incorporating friction-reducing additives in lubricant formulations require further investigation.

When considering what further research work should be done, it is apparent that the effect of lubricant ageing on fuel economy has not been studied in the past. As it is probable that the use of a lubricant which has thickened in service could cause increased fuel consumption, this aspect of the influence of engine lubricants on fuel economy should be investigated. This could be particularly relevant in view of the possibility of extending oil drain periods in order to effect economies.

10.4 Conclusions

(1) It seems feasible that measures taken to reduce internal friction in gasoline (and diesel) engines by modifying the crankcase lubricant could effect fuel economy improvements.

(2) Such improvements could be effected by the reduction of the viscosity of the lubricant base fluid or the incorporation of friction-reducing additives. The use of VI improvers that suffer large temporary shear losses could also effect economies.

(3) Fuel economy benefits would be greatest under cold-start short-trip running conditions. Under these circumstances benefits of up to 10% appear possible by, for example, replacing an SAE 30 grade oil by its multi-grade analogue, an SAE 10W/30 oil.

(4) Caution must be exercised so that steps taken to improve fuel economy by modifying the lubricant do not have a deleterious effect on other operating factors such as oil consumption, engine cleanliness or wear protection

References

1. C. F. Kettering. Motor design and fuel economy. *Ind. Eng. Chem.,* **17**, No. 11 (1925)
2. A. E. Cleveland and I. N. Bishop. *Engineering*, **189**, No. 4913 (1960) 824
3. J. A. Whitehouse and J. A. Metcalfe. The power losses of reciprocating internal combustion engines: some preliminary motoring tests on two small modern engines. *Motor Ind. Res. Assoc. Rep.*, No. 5 (1957)

4. J. A. Whitehouse and J. A. Metcalfe. The influence of lubricating oil on the power output and fuel consumption of modern petrol and compression ignition engines. *Motor Ind. Res. Assoc. Rep.*, No. 2 (1956)
5. J. C. Cree and J. G. Withers. All-season high performance oil. *Automob. Eng.*, **45**, No. 1 (1955) 21
6. T. J. Sheahan and W. S. Romig. Lubricant-related fuel savings in short-trip cold-weather service. *Soc. Automot. Eng. Pap.*, No. 750676 (1975)
7. W. B. Chamberlin and T. J. Sheahan. Automotive fuel savings through selected lubricants. *Soc. Automot. Eng. Pap.*, No. 750377 (1975)
8. E. D. Davison and M. L. Haviland. Lubricant viscosity effects on passenger-car fuel economy. *Soc. Automot. Eng. Pap.*, No. 750675 (1975)
9. J. C. Bell and M. A. Voisey. Some relationships between the viscometric properties of motor oils and performance in European engines. *Soc. Automot. Eng. Pap.*, No. 770378 (1977)
10. J. E. Bennington, D. E. Cole, P. J. Ghirla and R. K. Smith. Stable colloid additives for engine oils – potential improvement in fuel economy. *Soc. Automot. Eng. Pap.*, No. 750677 (1975)
11. T. J. Risdon and B. A. Gresty. An historical review of reduction in fuel consumption in United States and European engines with MoS_2. *Soc. Automot. Eng. Pap.*, No. 750674 (1975)
12. U.S. Department of Transportation, Federal Highway Administration. *Nationwide Personal Transportation Study* (1969)
13. T. C. Austin and K. H. Hellman. Passenger-car fuel economy as influenced by trip length. *Soc. Automot. Eng. Pap.*, No. 750004 (1975)

11 The Effect of Transmission Lubricants on Fuel Economy

E. L. PADMORE

11.1 Introduction

Very little work has been carried out on the effect of transmission lubricants on fuel economy. In fact most of the work reported has been done quite recently by the Lubrizol Corporation in the US. This means that the investigations into lubricant-related fuel savings have involved large American-built vehicles and not the generally smaller European models.

The findings may be valid for European vehicles but, at the present time, this cannot be substantiated. All the vehicles used for the test track investigations incorporated an automatic transmission. For clarity, we shall try to consider each component of the drive line separately (i.e. axle and automatic transmission), but the effect of different engine oils will also be considered because most of the work in this area considers all three together. Cumulative savings which can be achieved by selective lubrication will then be the sum of the savings obtained for each component.

11.2 Fuel Economy Related to Axle Lubrication (Vehicle Track Tests)

Gear-lubricant viscosity and performance-additive selection have been related to fuel economy by Sheahan and Romig[1] of Lubrizol. In all initial tests a Society of Automotive Engineers (SAE) grade 10W-40 engine oil (oil A) was used. The composition and physical properties of all of the oils tested are given in tables 11.1 and 11.2.

Figure 11.1 compares the results obtained using an SAE 80W-90 grade baseline gear lubricant (GA), an SAE 85W-140 lubricant (GB) and an SAE 80W-140 lubricant (GC). These data, which were obtained in side-by-side tests at an average ambient temperature of –2.8 °C (27 °F), represent multiple runs in each of the three cars with samples rotated to negate vehicle effects. As shown in table 11.3, the SAE 80W-140 lubricant, which contains the same performance additive as that in the SAE 85W-140 grade but is produced from a light-viscosity base stock and selected polymer, gave a 5%

Table 11.1 Composition and physical properties of engine oils

Oil[a]		A	B	C
Viscosity grade, SAE		10W-40	10W-40	10W-40
Base stock[b]		M	M	S
Base stock composition (Mineral oils)		40%v 100N 60%v 200N	80%v 100N 20%v 200N	–
Viscosity improver[c]		Y	T	Y
Physical properties				
Viscosity at 210 °F (99 °C),	cSt	15.33	14.59	14.94
	SUS	79	76	78
Viscosity at 100 °F (38 °C),	cSt	95.84	101.75	86.80
	SUS	444	472	402
Viscosity at 0 °F (–18 °C),	cP	23.40	24.00	12.20
VI		180	159	193

VI, viscosity index; SUS, Saybolt universal seconds (a viscosity measure).
[a]Same performance additive used in each oil.
[b]M, mineral oil (solvent-refined mid-continent stocks); S, synthetic hydrocarbon.
[c]Y, T, unspecified codes for type of viscosity improver.
(1 cSt = 1×10^{-6} m²/s; 1 cP = 1×10^{-3} N s/m²).

improvement over the SAE 80W-90 baseline lubricant in a 1 mile trip and a 1.2% improvement in a trip of 8 miles. Except in the first mile, the fuel economy achieved with the SAE 85W-140 grade is equivalent to that of the baseline grade; however, even the improvement at 1 mile is hardly significant (figure 11.1).

These results are rather surprising, because the advantage shown by 80W-140 oil occurs in spite of the fact that at high temperatures its viscosity is higher (28.18 cSt at 210 °F) than that of the baseline (15.48 cSt at 210 °F). Furthermore, at the low temperatures of the Brookfield test, which are near those of the cold-start fuel economy experiment, the viscosity of the 80W-140 is apparently the same as that of the baseline oil. No explanation is offered by the authors, and data on the lubricant temperatures and the viscosity of the base oil used in the 80W-140 oil are not given. A possible explanation may be that the effective viscosity of a gear lubricant containing large amounts of polymer is not given accurately by the SAE classification, i.e. there may be temporary viscosity loss under the high-stress conditions or alternatively a viscosity loss due to high shearing in such oils. Further experimentation is needed to elucidate such mechanisms more fully.

All the above data were obtained using full-sized American vehicles. Additional tests were conducted using what are termed 'compact' cars, with SAE 90 (GD) and SAE 75W (GE) gear lubricants and engine oil A. The results of these tests are given in table 11.4. They show clearly the benefit in fuel economy to be gained by using a lower-viscosity lubricant. A fuel

Table 11.2 Composition and physical properties of gear lubricants

Lubricant		GA	GB	GC	GD	GE	GF
Viscosity grade, SAE		80W-90	85W-140	80W-140	90	75W	75W
Base stock		Mineral oil	Mineral oil	Mineral oil	Mineral oil	Synthetic hydro-carbon	Synthetic hydro-carbon
Physical properties							
Viscosity at 210 °F (99 °C),	cSt	15.48	29.30	29.18	18.84	5.70	5.65
	SUS	80	139	134	94	45	45
Viscosity at 100 °F (38 °C),	cSt	169.60	475.11	344.47	233.60	33.82	33.50
	SUS	789	2202	1597	1083	158	157
Brookfield viscosity,	cP						
at 10 °F (–18 °C)		–	54000	–	–	–	–
at –15 °F (–26 °C)		144000	–	149000	–	–	–
at –40 °F (–40 °C)		–	–	–	–	83000	134000
VI		100100	95	120	98	120	119

1 cSt = 1 x 10^{-6} m^2/s; 1 cP = 1 x 10^{-3} N s/m^2.

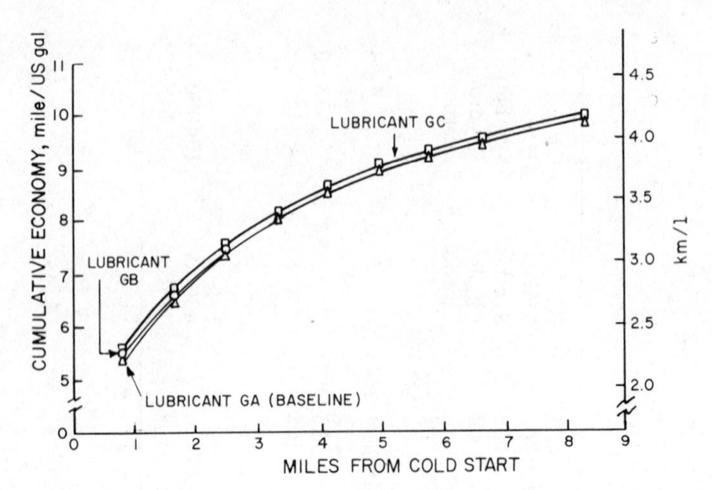

Figure 11.1 Effect of gear lubricant on cold-start fuel economy: full-size American cars[1] (1 mile/US gal = 0.425 km/l; 1 mile = 1.609 km)

economy improvement of 3.6% was shown for a 1 mile trip and an improvement of 1.0% for a trip of 8 miles. All these data were obtained at an average ambient temperature of –1.1 °C (30 °F).

In order to show the cumulative effect of differing engine and axle lubricants, a small series of tests were conducted using the lubricant combinations shown in table 11.5. The three lubricant sets were again compared in side-by-side tests, this time at an average ambient temperature of 1.7 °C

Table 11.3 *Effect of gear lubricant on cold-start fuel economy: full-sized American cars*

Trip length, mile	Cumulative fuel economy, from cold start, mile/US gal			Improvement in economy (GC versus GA), %
	SAE 80W-90 (GA)	SAE 85W-140 (GB)	SAE 80W-140 (GC)	
1	5.60	5.68	5.88	5.0
2	6.89	6.92	7.13	3.5
3	7.81	7.83	8.00	2.4
4	8.45	8.47	8.61	1.9
5	8.91	8.93	9.04	1.4
6	9.27	9.27	9.38	1.2
7	9.55	9.53	9.67	1.2
8	9.78	9.78	9.90	1.2

Average ambient temperature, 27 °F (–3 °C); engine oil, oil A.
1 mile/US gal = 0.425 km/l.

Table 11.4 Effect of gear lubricant on cold-start fuel economy: compact American cars

Trip length, mile	Cumulative fuel economy, from cold start, mile/US gal		Improvement in economy, %
	SAE 90 (GE)	SAE 75W (GF)	
1	8.36	8.66	3.6
2	10.00	10.24	2.4
3	11.04	11.25	1.9
4	11.74	11.92	1.5
5	12.24	12.40	1.3
6	12.63	12.78	1.2
7	12.93	13.07	1.1
8	13.17	13.30	1.0

Average ambient temperature, 30 °F (−1 °C); engine oil, oil A.
1 mile/US gal = 0.425 km/l.

(35 °F) using full-sized American vehicles. Lubricant set 1 (baseline) consisted of an SAE 10W-40 engine oil (oil B) and an SAE 80W-90 gear lubricant (GA); lubricant set 2 consisted of the same engine oil with an SAE 75W gear lubricant (GE); lubricant set 3 a synthetic-based SAE 10W-40 engine oil (oil C) and gear lubricant GE. The results of tests using these lubricant combinations are given in table 11.6. Lubricant set 2 showed a 2.5% increase in fuel economy over the baseline set in a 4 mile trip and a 2.4% increase in a trip of 8 miles. Lubricant set 3 gave 3.2% and 2.2% improvements over the baseline for trips of 4 miles and 8 miles respectively. These results, presented in figure 11.2, show the advantage to be gained by using lubricant set 3 as against lubricant set 2 for trips of between 2 and 5 miles. This benefit in fuel economy was not apparent, however, in trips greater than 5 miles.

The good fuel economy performance of lubricant set 3 was further confirmed in tests conducted in 'compact' vehicles, as shown in table 11.7.

Table 11.5 Lubricant sets for cold-start fuel economy tests

Lubricant set	SAE 10W-40 engine oil	Axle lubricant
1 (baseline)	B	GA (SAE 80W-90)
2	B	GD (SAE 75W)
3	C	GD

Table 11.6 Effect of lubricant set on cold-start fuel economy: full sized American cars

Trip length, mile	Cumulative fuel economy from cold-start, mile/US gal		
	Lubricant set 1	Lubricant set 2	Lubricant set 3
1	6.28	6.54	6.53
2	7.60	7.83	7.92
3	8.45	8.66	8.76
4	9.02	9.24	9.31
5	9.43	9.67	9.70
6	9.75	9.99	10.00
7	10.01	10.25	10.24
8	10.22	10.47	10.44

Average ambient temperature, 35 °F (2 °C).
1 mile/US gal = 0.425 km/l.

In these vehicles, set 3 gave a 3.0% improvement over the baseline set in a 4 mile trip and 2.7% improvement in a trip of 8 miles.

Sheahan and Romig[1] draw the following conclusions from their work.

(1) Short trips are highly significant as related to US driving patterns. Such trips consume more gasoline owing to warm-up factors and driving conditions.

(2) Selected engine oils and gear lubricants improved fuel economy by 5.0 to 0.9% in cold-start cold-weather service for trips of from 1 to 8 miles.

(3) Fuel economy improvements found with sets of selected engine oils and gear lubricants for a 4 mile trip from a cold start averaged 3.2 and 3.0% for full-sized and compact cars, respectively.

(4) Improvements found with selected engine oils or axle lubricants alone ranged from 1.1 to 2.5% for a cold-start trip of 4 miles.

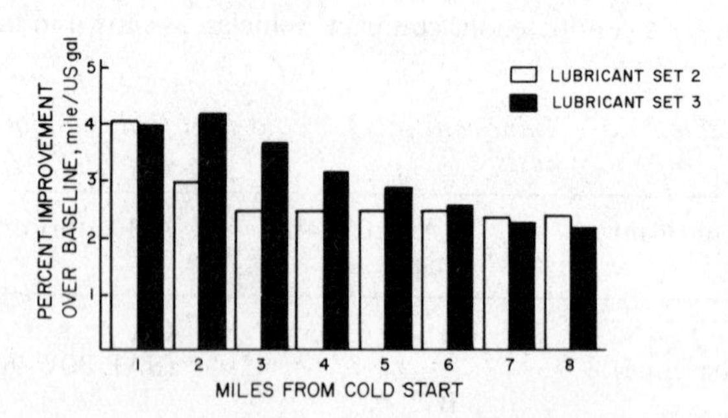

Figure 11.2 Lubricant sets 2 and 3 versus baseline set: full-sized American cars[1] (1 mile/US gal = 0.425 km/l; 1 mile = 1.609 km)

Table 11.7 Effect of lubricant set on cold-start fuel economy: compact American cars

Trip length, mile	Cumulative fuel economy from cold-start, mile/US gal		Improvement in economy, %
	Lubricant set 1	Lubricant set 3	
1	8.88	9.16	3.2
2	10.27	10.69	4.1
3	11.40	11.75	3.1
4	12.10	12.46	3.0
5	12.60	12.95	2.8
6	13.00	13.34	2.6
7	13.30	13.64	2.6
8	13.50	13.86	2.7

Average ambient temperature, 35 °F (2 °C).
1 mile/US gal = 0.425 km/l.

(5) Lubricant-related fuel economy improvements diminished with increasing trip length (from 1 to 8 miles).

They also made the following two conclusions concerning their test technique and the extremely important effect of ambient temperature.

(6) Statistical treatment of fuel economy data showed ambient temperature as a highly significant variable in cold-start short-trip cold-weather service.

(7) Coefficients developed by multiple linear regression analyses reflected an increase in fuel economy of 7.0% for full-sized cars and 7.6% for compact cars in a 4 mile trip for a 10 °F (6.3 °C) increase in temperature in the range 33–35 °F (1–2 °C).

11.3 Axle Efficiency Related to Lubricant Viscosity and Performance Additive Selection (Rig Tests)

Axle efficiency is a measure of the ability of the rear axle to convert input power into output power against friction from mechanical parts and lubricant viscosity. Relative axle efficiencies may be obtained, for various lubricants, by calculating an energy or heat balance for the axle. However, because of other driveline power losses from the transmission, wheel-bearing friction, tyre rolling resistance and universal joints, it is very difficult to relate improved axle efficiency to a finite improvement in vehicle fuel economy. In addition, axle efficiency varies with temperature, input speed and input torque. It follows therefore that laboratory tests run under prescribed

Table 11.8 Test conditions used in the rig test for the measurement of axle efficiency

Axle speed, rev/min	Power output, hp	Coolant flow rate, lb/min	Observed collant water temperature, °F	Observed lubricant temperature, °F
800	22.80	2	48–57	115–154
600	51.75	3	48–57	133–177
875	87.80	4	48–57	150–198

operating conditions can only give an indication of potential fuel economy gains in service.

The test apparatus set up for this investigation by Chamberlin and Sheahan[2] consisted of a standard hypoid axle fitted with a 6¾ in ring gear (4:1 gear reduction) but with a differential assembly removed. The axle was driven by a high-performance 454 cubic inch displacement (CID) (7.44 l) V8 engine via a four-speed manual transmission. The axle was enclosed in a metal box which allowed circulation of cooling water through the axle housing for temperature control. A well-run-in axle was filled with the test lubricant and was run for 5 minutes. The axle was then drained and refilled with a fresh charge of test lubricant. The test was run at three output torques

Table 11.9 Composition and physical properties

	GA (baseline)	GB	GC	GD
SAE viscosity grade	90	90	90	90
Base stock[a]	M [85%v 150B, 15%v 200N]	M [71%v 150B, 29%v 200N]	M [71%v 150B, 29%v 200N]	M [71%v 150B, 29%v 200N]
Viscosity improver	No	No	No	No
Physical properties				
Viscosity at 210 °F (99 °C), cSt	23.01	17.63	18.84	18.53
SUS	111	88	94	92
Viscosity at 100 °F (38 °C), cSt	322.2	215.0	223.6	234.6
SUS	1493	997	1083	1087
VI	97	96	98	95

[a]M, mineral oil (solvent-refined mid-continent stocks); S, synthetic hydrocarbon.
1 cSt = 1×10^{-6} m²/s.

and the coolant flow rate was held constant for each condition as outlined in table 11.8. At equilibrium lubricant temperature axle efficiencies were measured by heat rejection to the cooling water, under conditions approximating to 40, 55 and 70 mile/h road load.

Axle efficiencies were calculated using the following equations:

$$\text{efficiency} = \frac{\text{hp out}}{\text{hp in}} \times 100 \tag{11.1}$$

$$\text{hp in} = \text{hp out} + Q(\text{heat loss}) \tag{11.2}$$

$$Q(\text{heat loss}) = Q_{H_2O} + Q(\text{radiation and conduction}) \tag{11.3}$$

Assuming that Q(radiation and conduction) is negligible with the insulation of box,

$$Q(\text{heat loss}) = Q_{H_2O} \tag{11.4}$$

and thus

$$\text{efficiency} = \frac{\text{hp out}}{\text{hp out} + Q_{H_2O}} \times 100 \tag{11.5}$$

Also

$$Q_{H_2O} = WC_p\,\Delta T \tag{11.6}$$

where W is the water flow, lb/min, C_p is the specific heat of water,

aseline and candidate axle lubricants used in the rig test

;E	GF	GG	GH	GI	GJ	GK	GL	GM	GN
0	80W-90	80W	75W	75W	75W	75W-90	75W-90	140	80W-140
1 30%v 150B 70%v 200N	S	M [200N]	S	S	S	S	S	M	M
Jo	Yes	No	No	No	No	Yes	Yes	No	Yes
10.50	14.43	6.20	5.77	5.80	5.60	14.85	15.14	29.36	26.80
61	76	46	45	45	45	77	78	140	127
90.30	123.1	43.46	34.57	36.18	33.65	124.6	129.4	474.9	312.7
19	570	202	162	169	158	577	600	2201	1449
06	128	97	119	112	115	127	131	96	123

Table 11.10 Axle efficiency tests with varied cooling conditions

Axle lubricant		Operating speed, mile/h	(km/h)	Constant cooling to axle					Varied cooling to axle				
				Water in, lb/min	Water in, °F	Luboil temperature, °F	(°C)	Efficiency, %	Water in, lb/min	Water in, °F	Luboil temperature, °F	(°C)	Efficiency, %
GA (SAE 90 baseline)	Test 1	40	(64)	2.0	56	144	(62)	93.9	4.19	50	139	(59)	93.1
		55	(89)	3.0	53	169	(76)	95.7	6.38	47	165	(74)	94.8
		70	(113)	4.0	51	193	(89)	96.5	3.47	50	199	(93)	96.4
	Test 2	40	(64)	2.0	56	143	(62)	94.1	4.16	49	140	(60)	93.2
		55	(89)	3.0	52	168	(76)	95.8	6.53	48	166	(74)	95.2
		70	(113)	4.0	50	189	(87)	96.6	4.00	50	199	(93)	96.4
GC (SAE 90)		40	(64)	2.0	72	146	(63	95.2	2.39	52	140	(60)	93.9
		55	(89)	3.0	69	168	(76)	96.2	3.11	50	164	(73)	95.8
		70	(113)	4.0	67	189	(87)	97.1	1.76	57	201	(94)	96.8

Table 11.11 Axle efficiency screening tests with constant cooling to axle

Axle lubricant	Luboil temperature, °F, at			Axle efficiency, %, at		
	40 mile/h (64 km/h)	55 mile/h (89 km/h)	70 mile/h (113 km/h)	40 mile/h (64 km/h)	55 mile/h (89 km/h)	70 mile/h (113 km/h)
GA (SAE 90)	140	166	199	94.1	95.8	96.6
GB (SAE 90)	146	168	189	95.2	96.2	97.1
Repeat	140	162	183	95.4	96.3	97.0
GC (SAE 90)	145	163	185	95.6	96.6	97.2
GD (SAE 90)	140	163	183	95.8	96.8	97.2
GE (SAE 80)	127	150	168	95.8	96.7	97.3
GF (SAE 80W-90)	136	158	178	96.0	96.7	97.3
GG (SAE 80W)	126	147	168	95.8	96.8	97.1
GH (SAE 75W)	120	139	198	96.2	96.9	97.4
GI (SAE 75W)	121	139	154	96.0	97.1	97.4
GJ (SAE 75W)	115	133	150	96.8	97.5	97.7
GK (SAE 75W-90)	138	158	179	95.8	96.8	97.1
GL (SAE 75W-90)	137	157	175	96.1	96.8	97.3
GM (SAE 140)	154	177	197	94.8	96.0	97.0
GN (SAE 80W-140)	150	174	198	95.2	96.4	96.9

1 Btu/lb °F, and $\Delta T = T_{H_2O}$ out $- T_{H_2O}$ in. If $W = K$ lb/min and $\Delta T = \phi$ °F, then

$$Q_{H_2O} = K\phi \text{ Btu/min} \quad (11.7)$$

or

$$Q_{H_2O} = K\phi/42.4 \text{ hp} \quad (11.8)$$

and thus

$$\text{efficiency} = \frac{\text{hp out}}{\text{hp out} + K\phi/42.4} \times 100 \quad (11.9)$$

The composition and physical properties of the gear lubricants used in this investigation are shown in table 11.9. The baseline lubricant selected for the axle efficiency tests was a commercial SAE 90 grade gear oil of API GL-5 quality (GA) containing a sulphur–phosphorus additive.

The axle efficiencies for the SAE 90 grade baseline lubricant and another SAE 90 grade lubricant of slightly lower viscosity are shown in table 11.10. Two different sets of conditions were used in these tests. In the first case, constant cooling was applied to the axle, and the lubricant temperature was allowed to reach its own level; in the second case, a constant lubricant temperature was maintained by applying varied cooling to the axle. The results show that the lower-viscosity lubricant gave improved efficiency in both instances.

Since the second case more closely approximates to vehicle service, this condition was chosen for additional screening tests. Comparative data for all of the lubricants given in table 11.9, which vary in base stock composition and performance additive, are shown in table 11.11. It can be seen from table 11.11 that axle efficiencies increase with decreasing lubricant viscosity and changes in performance additive at conditions of 40 and 55 mile/h road load. At 70 mile/h, however, little variation in efficiency is apparent. The highest axle efficiencies were achieved with the SAE 75W synthetic base lubricant GJ.

It therefore follows that axle efficiency is affected by changes in lubricant viscosity and performance additive. A maximum improvement in efficiency of 1.8% was achieved (lubricant GJ), based on the average of the three sets of conditions.

11.4 Fuel Economy Related to Automatic Transmission Lubrication (Vehicle Track Tests)

Although this section is specifically concerned with automatic transmission lubrication, the effects of various combinations of engine oils, automatic transmission fluids (ATFs) and axle oils were determined during the

Table 11.12 *Composition and physical properties of engine oils*

Oil		C	I	J
SAE viscosity grade		40	10W-40	10W-40
Base stock[a]		M	M	S
Base stock Composition (mineral oils)		80%v 650N 20%v 350N	40%v 100N 60%v 200N	– –
Viscosity improver[b]			Y	Y
Physical properties				
Viscosity at 210 °F (99 °C),	cSt	14.45	14.94	12.94
	SUS	76	77	70
Viscosity at 100 °F (38 °C),	cSt	152.83	91.21	96.69
	SUS	708	423	448
Viscosity at 0 °F (–18 °C),	cP	17500	2080	1480
VI		101	183	142

[a] M, mineral oil (solvent-refined mid-continent stocks); S, synthetic hydrocarbon.
[b] Y, unspecified code for type of viscosity improver.
1 cSt = 1×10^{-6} m²/s; 1 cP = 1×10^{-3} N s/m².

investigation by Chamberlin and Sheahan[2]. The combinations chosen for the tests of automatic transmission lubrication were those that had been found to give the best results in dynamometer screening tests.

The composition and physical properties of the various engine and automatic transmission lubricants used in these tests are given in tables 11.12 and 11.13. Details of the axle lubricants are given in table 11.9. Again,

Table 11.13 *Composition and physical properties of automatic transmission fluids*

Fluid		TA (commercial baseline)	TD candidate
Base stock		M	S
Base stock composition (mineral oils)		90%v 100 neutral 10%v 200 neutral	–
Physical properties			
Specific gravity, 60°/60 °F		0.880	1.194
Viscosity at 210 °F (99 °C),	cSt	7.33	5.80
	SUS	50	45
Viscosity at 100 °F (38 °C),	cSt	41.13	40.41
	SUS	191	188
Brookfield viscosity,	cP		
at 0 °F (–18 °C)		1280	2500
at –40 °F (–40 °C)		37500	45750
VI		148	91

1 cSt = 1×10^{-6} m²/s; 1 cP = 1×10^{-3} N s/m².

Table 11.14 Track test lubricant sets

Lubricant set	Engine oil	ATF	Rear axle
1	SAE 40 baseline (C)	Commercial baseline (TA)	SAE 90 baseline (GA)
2	SAE 10W-40	TA	GA
3	SAE 10W-40 (J)	High density (TD)	SAE 75W (GJ)

lubricants were put into sets, and the three sets chosen are shown in table 11.14. Set 1 was the baseline set and consisted of an SAE 40 engine oil, a commercial automatic transmission fluid meeting 1975 car manufacturers' requirements and an American Petroleum Institute (API) GL-5 SAE 90 grade gear oil. Set 2 was modification of set 1, substituting a selected SAE 10W-40 engine oil for the SAE 40 grade lubricant. Set 3 was all synthetic based and consisted of an SAE 10W-40 engine oil, an experimental 'high-density' ATF and an SAE 75W synthetic gear lubricant.

The test procedure used for these tests was as follows.

The first ten laps of the federal durability driving schedule (eleven-lap sequence prescribed for complying with durability requirements of federal emission regulations) were selected as the basis for the driving cycle. In compliance with the nationwide speed limit, the maximum vehicle speed over this ten-lap cycle was 55 mile/h (89 km/h). Average vehicle speed was 29.5 mile/h (47.5 km/h). After break-in, three identical 1975 model full-sized American cars, powered by 350 in^3 (5.76 l) cubic inch displacement (CID) V8 engines, were started on test in caravan fashion but were spaced so as to avoid drafting. Three professional drivers and three lubricant sets were rotated among the three vehicles. Each day two ten-lap (75 mile) runs were made on the 7.5 mile track for a particular lubricant–car–driver combination. One run was made from a cold start, the other from a warm-engine condition. Fifty-four tests were required to rotate drivers, samples and cars fully and to make a.m. (cold-start) and p.m. (warm-engine) runs.

Fuel economy measurements were corrected to average ambient conditions for the track test. The equation for such corrections is given in the appendix to reference 2.

Table 11.15 shows corrected fuel economy data for the track test. It can be seen that lubricant set 2 gives a 2.2% improvement in fuel economy over set 1 (baseline lubricants) under cold-start conditions and a 1.1% improvement under warm-engine conditions. Lubricant set 3 gives a 1.3% improvement over baseline under cold-start conditions and a 0.5% improvement under warm-engine conditions.

Table 11.15 Fuel economy track test: corrected fuel economy[a]

Lubricant set	Time of test	Mean fuel economy, mile/US gal	Improvement over baseline	
			mile/US gal	%
1 (baseline)	a.m.	12.90	–	–
	p.m.	13.14	–	–
2 (engine oil 1/baseline ATF/ baseline gear oil)	a.m.	13.18	0.28	2.2
	p.m.	13.29	0.15	1.1
3 (engine oil J/ATF TD/gear oil GJ)	a.m.	13.07	0.17	1.3
	p.m.	13.20	0.06	0.5

[a]To average ambient conditions during test (30.76 in Hg pressure (104.2 kN/m^2), 52.6 °F (11.4 °C) and 46 grains/lb (6.57 g/kg) dry air humidity).

Linear regression analyses of the variables affecting fuel economy in these tests show that the increased fuel economy evident with lubricant sets 2 and 3 during p.m. runs (warm-engine conditions) could well be offset by driver variability even with professional drivers. It was therefore concluded that such incremental gains in fuel economy would not be meaningful in an everyday motoring situation, where far greater driver variability occurs. On the other hand, lubricant effects are much more significant than driver effects under cold-start conditions. Therefore, selected lubricants may give significant fuel savings in actual service under these conditions. (The same may be true of operation under warm-engine conditions, but this has not been proved in these tests.)

Lubricant set 3, which combines a synthetic-based SAE 10W-40 engine oil with a synthetic-based high-density ATF and a synthetic-based SAE 75W axle lubricant, was the one that gave the best results in the dynamometer screening tests. However, this was not the case under the conditions of the track tests. Additional tests were carried out to examine the interaction between the high-density ATF and the engine oil and rear-axle lubricant. Table 11.16 gives the results of preliminary fuel economy tests with the high-density ATF interchanged with the commercial ATF (lubricant set 4). Fuel economy was corrected to average ambient conditions for the first track tests (30.76 in Hg pressure (104.2 kN/m^2), 52.6 °F (11.4 °C) and 46 grains/lb dry air humidity (6.57 g/kg)) using equation 21 from the appendix to reference 2. In order to make these comparisons, a.m. and p.m. data for lubricant set 1, 2 and 3 from the original track tests were averaged. The few data for lubricant set 4 do not permit separation of a.m. and p.m. runs, and the improvement shown with this lubricant set can only serve as an indication of improved fuel economy.

Table 11.16 Fuel economy track test: preliminary results on high-density ATF effects (a.m./p.m. average)

Lubricant set	Corrected[a] fuel economy, mile/US gal (km/ℓ)	Improvement percentage over lubricant set		
		1	2	3
1 (baseline)	13.02 (5.53)	–	–	–
2 (engine oil L/baseline ATF/baseline gear oil)	13.24 (5.63)	1.7	–	0.8
3 (engine oil J/ATFTD/gear oil GJ)	13.13 (5.58)	0.8	–	–
4 (engine oil J/baseline ATF/gear oil GJ)	13.37 (5.68)	2.7	1.0	1.8

[a]To 30.76 in Hg (104.2 kN/m²) pressure, 52.6 °F (11.4 °C) and 46 grains/lb (6.57 g/kg) dry air humidity.

Comparison of the results for lubricant set 4 with those for set 3 suggests a decrease in fuel economy, under the conditions of the track test, using a high-density ATF. A substitution of the commercial fluid gave a 1.8% improvement in fuel economy. The reason for this unexpected result is not understood at this time. It could be caused by a complex interaction of throttle position, exhaust gas recirculation rate, spark advance and air/fuel ratio. The full fuel economy potential for the high-density ATF should be investigated with appropriate adjustments in other operating variables to take advantage of the improved efficiency of the transmission torque converter. A more direct coupling between engine and torque converter improves the efficiency of the latter and increases fuel economy; however, the change from a 'loose' to a 'tight' converter results in a loss of acceleration (assuming that the rear-axle ratio is not changed). In the track test the cars were operated in caravan fashion and comparable speeds, acceleration and deceleration times were maintained. Throttle positioning for equal acceleration may have obscured the benefit of the high-density ATF.

From this work Chamberlin and Sheahan draw the conclusion that vehicle track tests have demonstrated improved gasoline-engine fuel economy (from 0.5% to 2.7%) through selected combinations of engine oil and power-train lubricant.

A recent study of vehicle track fuel economy by Davison and Haviland[5] of General Motors showed broad agreement with the results of Lubrizol. The General Motors experiments were made with one full-sized US car and showed that fuel economy can be improved by reducing the viscosity of the engine and power-train lubricants. Using 'commercial' engine and axle lubricants, they were able to demonstrate a fuel economy difference between the high- and low-viscosity lubricants of 5% for a cold-start suburban cycle test (General Motors city–suburban cycle). They note that a reduction in engine oil viscosity gives the largest effect and that a reduction in axle lubricant viscosity has a larger effect than a reduction in transmission fluid viscosity. They also observed that in warmed-up conditions at constant speed, or for a highway driving cycle, a 2–3% fuel economy gain was shown for low-viscosity engine (SAE 10W) and rear-axle (SAE 80W) lubricants as compared with high-viscosity lubricants (SAE 20W-50 and SAE 90, respectively).

11.5 Work on Axle-lubricant-related Fuel Economy Carried out by Vehicle Manufacturers

There appear to be two routes to improved axle efficiency and associated fuel savings with gear oils. Firstly through the use of EP additives which improve the efficiency of the rear axle, and secondly through the use of lower-viscosity oils. One manufacturer[3] has obtained data indicating that under certain operating conditions passenger-car fuel consumption may be improved by as much as 5% by the simple substitution of a low-friction EP GL-5 additive for a conventional GL-5 additive.

Other studies have indicated that similar improvements could be obtained using rear-axle lubricants of much lower viscosity than those currently in use in conventional service applications. Specifically, lubricants meeting SAE 75W requirements offer improved efficiency as compared with conventional SAE 90 grade lubricants at a common operating temperature. While the use of less viscous axle lubricants may be viewed with some reservations, McClintock and Osborne[4] have demonstrated that lubricants in this viscosity range (approximately 4.3 cSt at 210 °F) will provide adequate gear and bearing protection when used in combination with high-quality additive systems of the type used to provide GL-6 performance. Obviously, the greatest degree of improvement in terms of axle efficiency and associated fuel economy will result when low-friction GL-5 additives are combined with low-viscosity base oils. Laboratory and field test evaluations of lubricants of this type are now in progress. Low-friction hypoid gear oils may well become the next major development in this market area.

11.6 Conclusions

(1) Selected combinations of engine oil and power-train lubricant can result in an overall improvement in fuel economy of the order of 0.9–5% in cold-start cold-weather service for trips of 1–8 miles.

(2) Lubricant-related fuel economy improvements diminish with increasing trip length from 1 to 8 miles.

(3) Fuel economy improvements after the first 8 miles, while small, are not negligible (up to 3%).

(4) The above applies to both full-sized and compact American cars. There are very few equivalent data available on European cars.

(5) Ambient temperature can significantly affect the apparent fuel economy improvement resulting from selected lubrication.

(6) There is some evidence to suggest that the effect of lubricant viscosity on fuel economy is greater for engine oils than for rear-axle lubricants which in turn is greater than for automatic transmission fluids.

(7) Only lubricants of a limited viscosity range have been tested so far: this could be extended provided there is a suitable choice of additives for maintaining lubricant performance.

References

1. T. J. Sheahan and W. S. Romig. Lubricant-related fuel savings in short-trip cold-weather service. *Soc. Automot. Eng. Pap.*, No. 750676 (1975)
2. W. B. Chamberlin and T. J. Sheahan. Automotive fuel savings through selected lubricants. *Soc. Automot. Eng. Pap.*, No. 750377 (1975)
3. L. F. Shiemann. Trends in hypoid-gear lubrication. *Natl. Lubr. Grease Inst. Spokesman* (April 1975) 15
   Also Paper presented at the *Natl. Lubr. Grease Inst. 42nd Ann. Meet., Chicago, Illinois, October 1974*
4. W. McClintock and R. E. Osborne. Effect of lubricant viscosity on rear-axle gear and pinion bearing operation. *Lubr. Eng.*, **20** (1974) 387
5. E. D. Davison and M. L. Haviland. Lubricant effects on passenger-car fuel economy. *Soc. Automot. Eng. Pap.*, No. 750675 (1975)

12 Mileage Marathons

W. S. AFFLECK and G. B. TOFT

12.1 Introduction

It must be accepted that the motor car is not an outstandingly efficient device for converting chemical energy into useful mechanical work. The question of why efficiencies are characteristically so low can be discussed at many levels, but at a strictly practical level it reduces to how far *could* a motor car go on a quantity of fuel if some or most of the constraints on its construction and use were removed. This question forms the basis of the mileage marathons which have been run at the Shell laboratories for a number of years.

12.2 History

The start of the Shell mileage marathon at Shell Oil's Wood River Laboratory in 1939 stemmed from an argument amongst engineers about ultimate fuel economy. The first event was won by R. J. Greenshields with a run of 49.73 mile/US gal (4.73 l/100 km). Ten years later, Greenshields ran a modified Studebaker to 149.95 mile/US gal (1.57 l/100 km)[1] (plate 12.1). In 1968 a Fiat 600 owned by J. M. Jones, R. C. Trokey and D. C. Carlson reached 244.35 mile/US gal (0.963 1/100 km)[2] (plate 12.2) and in 1973 an Opel driven by B. E. Visser achieved 376.59 mile/US gal (0.625 1/100 km)[3] (plate 12.3). Over the years many aspects of the marathon have changed: initially, cars were run until a fixed amount of fuel was consumed and the distance to run-out recorded. Traffic and safety considerations caused a change to a procedure where the fuel used over a fixed course was the crucial measurement. At one stage the event was run to strict average speed timings over a country road course, but, later, traffic and safety conditions forced a further change. Latterly the event was run on a relatively level course over a divided highway near the laboratory. Throughout the competition, vehicles have been required to bear a clear lineal relationship to some production car. The minimum quoted weight for the parent model and the use of an engine available in that model have restricted development – although, to judge from the modifications noted in references 2 and 3, considerable scope remained for improvement in performance within these restrictions.

Research staff at the Thornton Research Centre (TRC) ran the first Shell mileage marathon in Europe in 1969. The event was modelled on the Wood

Plate 12.1 A modified 1947 Studebaker which achieved 149.95 mile/US gal in the 1949 Wood River competition

Plate 12.2 A modified 1959 Fiat 600 which achieved 244.35 mile/US gal in the 1968 Wood River competition

Plate 12.3 A modified 1959 Opel which achieved 376.59 mile/US gal in the 1973 Wood River competition

River class for slightly modified production cars. Driving style was not restricted, but the extent to which a normal production car could be tuned was limited to changes in carburation and ignition timing. The event was run on a closed airfield circuit with a minimum average speed of 30 mile/h (48 km/h) enforced. Awards for best miles per gallon and best ton miles per gallon (miles per gallon multiplied by vehicle weight) have been made. Small cars (Fiat 500s, Imps, Minis) have managed about 90 mile/gal (3.1 l/100 km) under these conditions, the current best being 96 mile/gal (2.94 l/100 km) by a Mini 1000 driven by B. D. Caddock. Larger cars, notably British Leyland 1800s, have given best ton miles per gallon results, 109.0 ton mile/gal having been achieved by I. C. H. Robinson. These impressive values were due more to the driving technique employed than to engine tuning.

In 1973 a competition for special vehicles was initiated. These vehicles were required only to be genuinely two-track vehicles and a classic three-wheel configuration has evolved[4]. Competition was over a 10 mile (16 km) course with a speed minimum of 10 mile/h (16 km/h) average enforced. In 1976 a special (see later) driven by B. W. Beattie achieved 1141 mile/gal (0.248 l/100 km)*.

12.3 Some Theory

How is it done? Marathoning is dominated by two considerations. Firstly the power needed to propel the vehicle must be kept to an absolute minimum, and secondly the engine and operating conditions must be chosen so that that power requirement is met with minimum fuel utilization.

*The understanding of this chapter is not greatly effected by the choice of units, and so traditional units, which are still used by most competitors, are preserved. Conversion constants to other systems can be found in the appendix.

The power needed to drive a vehicle is given by the equation (see section 2.5.2)

$$\text{bhp} = \frac{V}{375}(C_R W + 0.0026 C_D A V^2 + GW + FW) \tag{12.1}$$

where bhp is the brake horsepower, V the speed, mile/h, C_R the coefficient of rolling resistance, W the vehicle weight, lb, C_D the aerodynamic drag coefficient, A the frontal area, ft^2, G the road gradient, fractional, and F the acceleration, fraction of g.

For a four-stroke engine, the available power is given by

$$\text{bhp} = 0.077 \times \text{bmep} \times C \times R \tag{12.2}$$

where bmep is the brake mean effective pressure, lb/in^2, C the swept volume of the engine, l, and $R \times 10^3$ the engine speed, rev/min.

The rate of fuel usage is given by the product of bhp and brake specific fuel consumption (bsfc). Although the bsfc varies over the whole range of engine speed and load conditions the effect of throttling at constant speed can be empirically approximated by the equation

$$\text{bsfc} = \frac{B_m \times \text{bmep}_m \times \text{imep}}{\text{imep}_m \times \text{bmep}} \tag{12.3}$$

where B_m is the bsfc minimum at bmep = bmep$_m$ and imep = imep$_m$. imep is the indicated mean effective pressure equal to bmep plus friction mean effective pressure (fmep). For approximate calculation fmep can be assumed independent of bmep (see further).

12.4 Minimizing Power Requirement

Level-road constant-speed power requirement is dominated by vehicle speed, and reducing speed is an obvious route to economy. The imposition of minimum speed requirement limits progress in this direction as can be seen if typical values* are taken for other quantities in the equation 12.1:

at 10 mile/h, bhp = 0.0267(40 + 2.08) = 1.12

at 30 mile/h, bhp = 0.08(40 + 18.72) = 4.7

at 60 mile/h, bhp = 0.16(40 + 74.88) = 18.38

Assuming that these power requirements could all be met at a bsfc of 0.08 gal/bhp h (≡ 0.6 lb/bhp h), level-road fuel economies of 111.6, 79.8 and 40.8 mile/gal would be predicted for these three speeds. These calculations show that, at low speed, aerodynamic drag is not very significant, rolling resistance being the dominant constraint.

*C_R = 0.02, C_D = 0.4, A = 20 ft^2, W = 2000 lb; C_R and C_D are rounded values of those given in section 2.5.2 and are used for baseline purposes in subsequent calculations.

12.4.1 Rolling Resistance

In practice it is easy for C_R to have an inflated value through brake binding, through tyre scuffing caused by inaccurate wheel alignment or as a result of drive line friction caused by faulty gear assembly or inadequate lubrication. It may be assumed that eliminating these fault conditions is a prerequisite to any form of economical motoring and a natural first step in preparing a car for marathon running.

Rolling resistance is, in a properly prepared vehicle, largely determined by tyre size, profile, structure and inflation pressure. Increasing inflation pressure on a conventional cross-ply 14 in tyre from 25 to 40 lb/in^2 reduces C_R from 0.018 to 0.015. Changing from cross-ply to a radial-ply tyre, inflated to 40 lb/in^2, will further reduce C_R to 0.011. Investigation with the Wood River Fiat 600 showed further reductions in rolling resistance as tyre pressure was increased to 100 lb/in^2. Increasing from 20 to 100 lb/in^2 increased the 16.5 to 0 mile/h coast distance by 50%.

Large wheels show useful reductions in rolling resistance over small ones, thin sectioned tyres over thick sectioned ones, and treadless tyres over fully treaded ones. TRC specials have used high-pressure racing bicycle tyres and probably achieve values of C_R of 0.003.

For a 'modified' car a reduction in C_R to 0.01 appears plausible, with a significant further reduction available to the special builder. Recalculating equation 12.1 with these data for a 2000 lb vehicle,

$$\text{at 10 mile/h, bhp} = 0.0267(20 + 2.08) = 0.59$$
$$\text{fuel economy at 0.08 bsfc} = 212 \text{ mile/gal}$$

$$\text{at 30 mile/h, bhp} = 0.08(20 + 18.72) = 3.09$$
$$\text{fuel economy at 0.08 bsfc} = 121.0 \text{ mile/gal}$$

12.4.2 Weight Reduction

Significant reductions in the weight of a production vehicle are impractical, but for the special vehicle weight reduction is a logical move. Calculating for a 500 lb vehicle with a C_R of 0.003 indicates that the 10 mile/h power requirement is

$$0.0267(1.5 + 2.08) = 0.095 \text{ hp}$$

and the predicted economy assuming again an available 0.08 gal/hp h bsfc is 1308 mile/gal.

12.4.3 Aerodynamics

It will have been noted that the effect of reducing weight and rolling resistance has been to elevate aerodynamic drag to a point of significance even at

10 mile/h for the special. For the special, a reduction of frontal area and some streamlining could plausibly reduce the product $C_D \times A$ to 3.0. Probably the potential for further reduction is limited since reducing frontal area further without increasing specific resistance is not indefinitely possible.

Calculating for 500 lb weight, $C_R = 0.003$ and $C_D \times A = 3.0$, the 10 mile/h requirement is

$$0.0167(1.5 + 0.78) = 0.061 \text{ hp}$$

or a theoretical 2053 mile/gal.

A 20% reduction in aerodynamic drag for the 2000 lb vehicle at 30 mile/h reduces the power requirement from 3.09 to 2.8 hp and increases economy from 121 to 134 mile/gal.

12.4.4 Performance on a Road

The modest values of power requirement and dramatic economy numbers derived in the previous sections all assume steady-state level-road conditions. In reality the vehicles must be able to accelerate and to surmount gradients. The Appleton circuit, for example, on which the TRC mileage marathon has been held approximates to a three-element course, 62% level road, 14.5% uphill at a mean gradient of 1.78% and 23.5% downhill at a mean gradient of 1.1%. The total circuit length is approximately 2 miles. Applying equation 12.1 to the case of the 500 lb special it is clear that the vehicle will free wheel down the 1.1% downgrade; indeed, its terminal velocity here would be over 20 mile/h. Equally it is clear that, if the engine is directly coupled, i.e. bhp $\leqslant$ constant x road speed, the vehicle will be incapable of climbing the 1.78% upgrade: indeed 0.061 bhp, if available through suitable gearing, would only maintain just over 2 mile/h on the upgrade. Approximate calculation indicates that exposing the 500 lb special to this course with only 0.061 bhp available would not allow a circuit average in excess of 8 mile/h to be maintained. More power is required. The ability of the vehicle to cover almost a quarter of the circuit distance under gravity means, however, that installed power can be increased by nearly a third without any sacrifice in fuel economy over the circuit.

Free wheeling without engine operation is a feature of mileage marathon driving technique which produces, where it is allowed by competition rules, a very substantial increase in economy. For legal and safety reasons it cannot be recommended to the everyday motorist. The rationale of the 'accelerate and coast' marathon technique is discussed in section 12.5.1.

The impact of acceleration needs and gradient climbing capability on power requirements has been discussed; no less significant for the achievement of ultimate performance are the factors of tyre scuff on cornering (minimized by adopting wide radius lines through curves) and of course the application of braking, for example, safely to enter a tight curve. The movement of a damped car suspension system also absorbs energy which must ultimately show up as reduced fuel economy. A smooth surface, where it can be selected by the driver will help, as will mechanical modifications to limit suspension travel. This is particularly necessary where suspension movement, because of the suspension geometry, involves tyre scuffing[3]. A lightweight special will normally be unsprung.

12.4.5 Transmission Efficiencies

Manual shift gearboxes and back-axle/differential gears on current production vehicles are quite efficient components. Efficiency (in terms of minimum energy loss in the transmission) can be improved by the use of low-viscosity lubricants, by the reduction of grease fillings and by the removal of oil/grease seals. Needless to say, none of these procedures could be recommended for long-term application in normal motoring. The effect of removing oil seals, wheel-bearing grease and conventional crankcase and gearbox oils, replacing the latter with light machine oils, was studied for the 150 mile/US gal 1947 Studebaker[1]. At 40 mile/h constant-speed running these measures saved 1.4×10^{-3} US gal/mile. The effect of the total vehicle preparation schedule was to reduce a requirement of 43.3×10^{-3} US gal/mile to 17.5×10^{-3} US gal/mile.

Specials and highly modified 'unlimited' marathon vehicles have used minimal transmission involving only a clutch, chain drive and often a free wheeling clutch which eliminated drive line drag during free wheel running. The efficiencies here are high enough to justify disregarding transmission losses in calculation.

In practical vehicles, power loss in automatic transmission is significant. Mileage marathon rules provide a handicap correction, usually of 8%, to compensate the owners of vehicles so equipped.

12.5 Providing the Power

Although the level-road load power requirements calculated for marathon vehicles have been shown to be unrealistically low and to require enhancement, even the uprated power need is very modest. Assuming that the engine will be operated at a bmep of about 100 lb/in^2 * at a practical minimum engine speed of 1500 rev/min, equation 12.2 indicates a specific power output of 11.55 bhp/l swept volume. The power needs of the 500 lb special could therefore be met by an engine of perhaps 15 cm capacity (assuming a three-fold increase over calculated level-road power is needed). 300 cm^3 would be an appropriate capacity for a 2000 lb marathon car.

*It is necessary to postulate a high bmep figure to make plausible a good brake specific fuel consumption value (see section 2.4).

For the special builder the availability of small-displacement moped engines provides a route to power requirement matching, but for the entrant whose vehicle must be powered by the original engine the problem is intractable. Attempts to run multi-cylinder engines with two or more pistons removed and appropriate valve and manifolding modifications have not produced spectacular fuel economy gains.

In normal road running a low-power requirement is met by operating the engine at part throttle, that is to say at low bmep. To drive a 2000 lb vehicle at 30 mile/h level-road load with mileage marathon preparation has been calculated to require 2.8 bhp. If this is supplied by a 1500 cm^3 engine driving at a typical effective gear ratio of 17 (mile/h)/(1000 rev/min) the bmep needed is, by equation 11.2, 13.7 lb/in^2. If it is further assumed that the best bsfc for that engine is 0.08 gal/bhp h at a bmep of 100 lb/in^2 and that the fmep for the engine is 18 lb/in^2, the expected bsfc at 13.7 lb/in^2 bmep is, from equation 11.3, 0.156 gal/bhp h. Engine efficiency has been almost halved by throttling back to match road-load power requirement, and the calculated fuel economy reduced from 134 mile/gal to 68 mile/gal.

12.5.1 Mileage Marathon Driving

Evidently a great problem in the practical realization of the excellent fuel economy figures predicted by road-load power requirement calculation is the utilization of the vast excess of engine power available in practical vehicles. The practical solution used whenever competition rules allow is to adopt a power-burst driving technique in which wide-open throttle (WOT) accelerations are coupled with engine-off coast sequences to achieve the required average speed.

Equation 12.1 shows that, on a level road, a marathon-prepared 2000 lb vehicle will decelerate at 0.38 mile/h s if set to free wheel at 30 mile/h. Assuming this rate of retardation is constant over the speed range 32.5–27.5 mile/h will occupy 0.017 mile. The total accelerate–coast cycle thus covers would take 13.1 s and would cover just over 1/10 mile. A 1500 cm engine geared as before and operating at 100 lb/in^2 bmep will give the vehicle an acceleration of 2.39 mile/h s at 30 mile/h. Accelerating from 27.5 to 32.5 mile/h will occupy 0.017 mile. The total accelerate–coast cycle thus covers 0.126 mile during only 13.7% of which is the vehicle under power. The engine is working at 20 bhp in an efficient mode where 0.08 gal/bhp h can be assumed as a bsfc value. The cycle requires 0.00094 gal, equivalent to 133.5 mile/gal. This figure may be compared with 68 mile/gal calculated for part-throttle cruising and 134 mile/gal for the 2000 lb vehicle with a marathon-power-matched engine. The excellent economy of the power-burst driving cycle is dependent on the engine-off coast; to leave the engine idling would probably increase the fuel requirement for the cycle by some 0.00025 gal and reduce economy to 105 mile/gal.

The example has been calculated with a 5 mile/h speed excursion to justify simple linear acceleration assumptions. In practice, speed would be raised over a wider range so as to minimize the number of engine start-ups required for each mile and to allow the accelerate–coast cycles to be phased to the topography of the course.

The accelerate–coast driving cycle is by far the most potent tool available to the marathon driver. It can, of course, hardly be applied to normal highway driving.

12.5.2 Choice of Vehicle

It is interesting to enquire whether the calculation of fuel requirement in the drive–coast cycle indicates any preferred power–weight characteristic for a marathon vehicle. To explore this question, the following calculations were carried out.

A 2000 lb vehicle required to average 30 mile/h over a level road was powered by 500 cm^3, 1, 2 and 4 l engines each capable of 0.08 gal/bhp h at 100 lb/in^2 bmep and geared to 17 (mile/h)/(1000 rev/min). The fuel used for acceleration–coast cycle decreased with engine capacity from 0.00149 gal for the 500 cm^3 engine to 0.00086 for the 4 l engine. However, because of the much sharper acceleration available with the large engine, the number of cycles required to cover a mile increased from 5.5 (for the 500 cm^3 engine) to 9.6 for the 4 l engine, and the sample calculation predicts that each vehicle will achieve essentially the same fuel economy. Practically, some fuel must be used simply to start the engine at each cycle, and this represents an incremental burden. If it is assumed that this represents 0.5 ml of fuel per start-up, the economy of the larger engines is reduced. With this assumption the 500 cm^3 engine should allow 113 mile/gal against 107 mile/gal for the 4 l engine. Since it is difficult to see less fuel being wasted in starting a large engine than a small one, a larger discrimination in favour of the smaller engine is likely.

Next a 2 l engine was coupled to vehicle weights of 1500 to 5000 lb, the aerodynamic drag ($C_D \times A$) being assumed constant. The fuel required per cycle increased with vehicle weight and the number of cycles per mile decreased. The effects were not cancelling, and the light vehicle achieved, on paper, 139 mile/gal against 68 mile/gal for the 5000 lb car. Calculated ton miles per gallon, however, clearly favoured the heavier vehicle, and this discrimination was increased when allowance was made for fuel lost in engine start-up. The 1500 lb vehicle was calculated at 80 ton mile/gal against 146 ton mile/gal for the 5000 lb car.

Small engines in small vehicles should yield best economy (predictably), while for ton miles per gallon competition the heavier the vehicle the better. In practice competition rules usually limit the extent to which a vehicle can be ballasted.

12.5.3 Engine Modifications

Production engines are designed to provide a compromise of performance, economy, emissions, weight, longevity and cost. There is no reason to suppose that this compromise cannot be shifted to favour economy at the expense of other factors.

Modern engines are required to drive a wealth of power-consuming accessories, the total burden of which can far exceed the marathon road-load requirement (see section 2.5.4). Depending on the rules governing competition, some or all accessories will likely be removed for marathon driving. The 150 mile/US gal Studebaker had its fuel requirement reduced by 0.0024 US gal/mile when accessories were removed; this was almost twice the benefit achieved by the lubrication changes cited previously and the 1947 automobile was sparsely equipped by modern standards. Since the bsfc values used in calculation heretofore reflect 'bare engine' values, it is not legitimate to claim any benefit in subsequent calculations for disconnecting accessories.

12.5.4 Carburation

Three features of carburetter design contribute disproportionately to fuel usage. These are the choke, which will not anyway be needed in marathon driving, the power-enrichment system and the accelerator pumps. Commonly all these three are disconnected for marathon driving, and the car is run entirely on the carburetter main metering system. Since fuel loss in start-up and, more important, excessive fuel loss in reluctant start-up contribute to depressing economy, it is necessary to compromise simplified carburation in the interests of startability. Again the changes suggested here would not affect the steady-state bsfc values used in calculation and credit cannot be taken for them.

It is well known that mixture strength affects both power output and bsfc. Typically a carburetted engine might show a maximum power at an air/fuel ratio of 13/1, a minimum bsfc at 16 or 17/1 (see figure 6.14). Bmep at maximum power might be 120 lb/in^2 reducing to 105 lb/in^2 at minimum bsfc. The corresponding bsfc values would be 0.076 and 0.069 gal/bhp h. Although the accelerate–coast marathon technique makes full use of available engine power, experience (and calculations) show that the engine can be leaned out to minimum bsfc with advantage. Applying the figures quoted above to a 2000 lb/1500 cm vehicle in 30 mile/h level-road marathon driving and neglecting fuel loss on start-up, it is found that, at 120 lb/in^2 bmep, time of use of the engine in the cycle is 13.1% and fuel economy 125 mile/gal. Leaning out the carburetter to the setting corresponding to 105 lb/in^2 bmep increases engine use to 14.4% and increases economy to 139 mile/gal.

The shape of the bmep–fuel flow 'fish hooks' reflects an engine's ability to run lean. This ability is increased substantially if the air–fuel mixture is fully vaporized in the inlet system. Since the process of vaporization reduces volumetric efficiency, there is a power loss associated with the change from carburetted to vaporized fuel. Typically the 105 lb/in^2 bmep of the lean carburetted engine might be reduced to 94 lb/in^2 with full fuel vaporization. However, the decrease in minimum bsfc, from 0.069 to 0.064 gal/bhp h, is found to more than offset this power loss. Calculations show an increase in engine use from 14.4% to 16.2% but an increase in economy from 139 to 151 mile/gal.

In the example calculated above and based on figure 6.14, the air/fuel ratio for the bsfc minimum was only shifted by about one number in converting from carburetted to vaporized fuel. Factors which influence the location of this minimum are the degree of charge heating needed to secure vaporization and the quality of the ignition system. To allow very lean operation in a marathon vehicle, high-energy ignition systems are commonly fitted.

12.5.5 Engine Heat Management

Because the marathon vehicle engine is only operated part of the time and because it is only operated in high-thermal-efficiency modes, the normal cooling arrangements for the engine will usually be excessive. It is important to keep the engine hot to minimize heat rejection from the combustion chambers, to ensure good fuel vaporization and to keep lubricant viscosity low. Elimination of the cooling fan, the provision of radiator masks, insulation of the engine structure and the re-cycling of exhaust heat to inlet system or to the sump are tried expedients in heat retention. Measurements on the effect of marathon preparation modifications have normally been carried out under constant-speed conditions, and hence the quantitative significance of these heat management techniques has not been determined.

12.5.6 Compression Ratio

An increase in compression ratio is a well-documented route to increased power and improved economy. Engine preparation for a marathon vehicle frequently involves machining the cylinder head or changing head gaskets to increase compression ratio. This is possible because the fuel supplied for the marathon is normally 'best' quality and because the engines can be tuned to take full advantage of, probably, a higher octane quality than their design norms. The accelerate–coast mileage marathon driving technique does represent a close approximation to normal knock rating, and the extent to which compression ratio can be raised will be limited by desires for engine preservation. Suppose, however, that the engine has a normal requirement of 92 octane number corresponding to 7.7:1 compression ratio. If the engine is now developed fully to utilize 100 octane number fuel, the compression ratio might be increased to 9.5:1 (see figure 3.6). The effect of this would be to increase bmep from 93 lb/in^2 at a bsfc of 0.07 gal/bhp h to 103 lb/in^2 at a bsfc of 0.06 gal/bhp h (see figure 3.1). Applying these figures to the 2000 lb/1500 cm^3 marathon vehicle operating in the 27.5/32.5 mile/h cycle, an improvement in fuel economy from 138 mile/gal to 161 mile/gal is calculated. It should be noted that the engine characterized in figure 3.1 is less efficient than that considered in earlier calculations, and hence the increase in economy computed above cannot be regarded as cumulative to values calculated previously.

Changing the compression ratio of the 1947 Studebaker from 6.5:1 to 10.0:1 was worth 2.4×10^{-3} US gal/mile at 40 mile/h. Greenshields[1] notes that the fuel economy benefit of a major change in engine performance of this kind can only be realized if transmission ratios are adjusted as well.

12.5.7 Engine Friction

The reason for the poor bsfc values obtained with lightly loaded engines is that at decreasing loads an increasingly high percentage of the power output of the engine is lost to overcoming engine friction and pumping work. Typically these may represent about 20 lb/in^2 mep, see figure 3.13. Ricardo[6] established a breakdown of frictional and pumping losses for a large low-compression aircraft-type gasoline engine where losses totalled 18 lb/in^2 fmep. The results are probably acceptable to an automotive engine. Fmep was divided between pumping losses (3.5 lb/in^2 mep), valve mechanism friction (2.5 lb/in^2 mep), crankshaft bearings (4.0 lb/in^2 mep) and piston friction (8.0 lb/in^2 mep). While at part-throttle pumping mep increased, there was a compensating drop in crankshaft bearing and piston friction so that the assumption of constant total fmep used in equation 12.3 is approximately valid.

For the dedicated marathon engine builder these figures highlight the scope for reduction in fmep. Larger valves, increased valve lift, port polishing and inlet and exhaust flow de-bottlenecking should help reduce the pumping component, perhaps by 0.5 lb/in^2 mep. Because the engine is not required to operate at high speeds, some adjustment to cam profiles and valve spring tensions may be admissible, perhaps reducing fmep by a further 0.25 lb/in^2. Crankshaft-bearing losses are sensitive to clearance, and precise bearing alignment and careful lapping and assembly could save 1 lb/in^2 mep here. Since oil control, crankcase blowby and piston cooling are not of paramount importance in marathon driving, part of the piston ring pack and piston clearance may be sacrificed to save, perhaps, a further 2 lb/in^2 mep. Totally these savings could reduce fmep from 18 to 14.5 lb/in^2.

Applying these savings (on paper) to the 1.5 l engine, 94 lb/in^2 bmep should be increased to 97.75; bsfc reduced from 0.064 to 0.0615 gal/bhp h. In the 2000 lb vehicle these changes could increase 151 mile/gal to 157 mile/gal.

12.6 Theory and Practice

The very simple theoretical approach developed in the previous sections has been given to establish the scale of effects possible when a vehicle is deliberately prepared and driven for maximum economy. The mileage marathon competitions remain essentially practical: roads are not straight and level; winds blow and rains fall. For many contestants the marathon vehicle is also the family car, and extensive modification is not acceptable. The organizers of the Wood River mileage marathons identified 'uncanny skill, superior knowledge and slick cunning' as essential attributes of the successful marathon contestant. It would be wrong to underestimate the importance of the first and last of these.

12.7 The Tale of a Successful 'Special'

The 1976 Thornton mileage marathon (special category) was won at 1141 mile/gal by a three-wheeled special built by a team of four Thornton employees (the late E. W. Ball, B. W. Beattie, E. R. Hinde and G. B. Toft) using mostly secondhand parts.

12.7.1 Thornton Research Centre Marathon Conditions and Rules

The marathon was held in June on the Appleton test track near Warrington. Competitors had to complete 10 miles on a 1.89 mile circuit in 1 h or less. Details of the old airfield circuit have been given earlier in this chapter. The fuel consumption was measured volumetrically to the nearest 1/20 ml. The rules of this particular competition define a 'special' as a vehicle with at least three wheels, each of which must support at least 20% of the total weight of the vehicle plus driver in his normal driving position. The power unit must be a four-stroke gasoline engine with no restriction on capacity or number of cylinders. The vehicle must have at least one wheel braking. The only other rule is that adjustable aids to forward motion are banned. Thus the vehicle can only be powered by energy obtained from the gasoline carried in the regulation glass fuel tank. The rules do allow the vehicle to coast with the engine switched off.

12.7.2 The Chassis

The chassis shown in plate 12.4 consists of a sports bicycle frame (obtained from a rubbish tip) with a simple space-frame outrigger attached to the nearside. This outrigger frame is built from small section 18 gauge mild steel tubes with a 20 gauge aluminium deck riveted on to support the driver who lies in the prone position. The driver's legs are supported by an extension of the outrigger frame. The chassis is carried on three bicycle wheels 26 in in

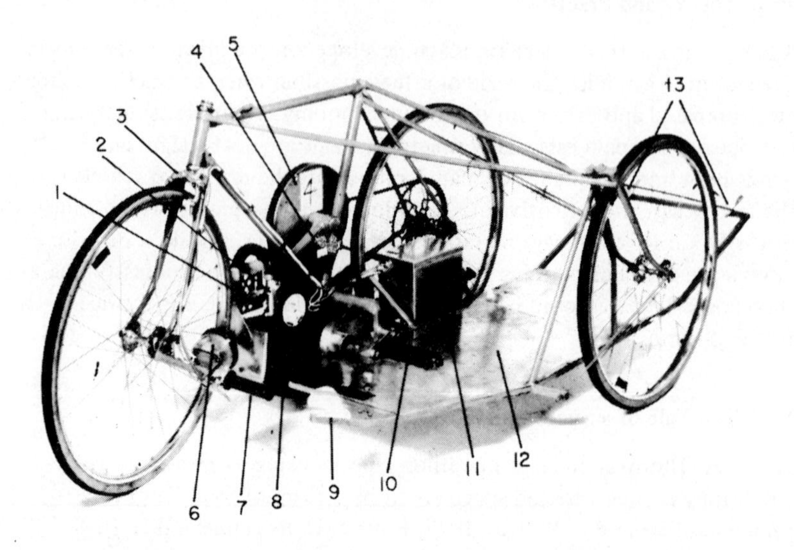

Plate 12.4 The 1976 Thornton mileage marathon winning special: 1, glass fuel tank and levelling mechanism; 2, special carburetter; 3, brake; 4, cylinder temperature indicator; 5, primary chain case; 6, tachometer; 7, twist grip ignition switch and brake lever; 8, engine insulation; 9, arm rest; 10, starter motor; 11, coil and battery box; 12, driver's platform; 13, leg rests

diameter with light alloy hubs and rims, and these are fitted with high-pressure (70 lb/in^2) racing tyres. The ball bearing hubs are lubricated by a very light oil, and the driven wheel incorporates a standard bicycle free wheel unit so that when coasting there is no transmission drag. The use of these large-diameter wheels with their high load bearing/weight ratio and very low rolling resistance due to the small contact area with the road allows the vehicle to coast for long periods of time, particularly on the downhill sections of the track.

Even though the test track is very rough in places the use of large-diameter wheels overcomes the necessity for any form of suspension, although some springiness is given by the front forks of the bicycle frame and the second pair of front forks that are attached to the outrigger frame to carry the 'sidecar' wheel. Steering is by a tiller attached to the front forks at hub level.

12.7.3 The Engine

The engine is a 49 cm^3 Honda OHV single-cylinder air-cooled four-stroke unit, type C100. This has a bore of 40 mm and a stroke of 39 mm. The compression ratio is 8.5:1, and this was not increased for the event. The rotating mass of the engine and gearbox, which are integral, was reduced by removing the magneto flywheel and by converting to 6 V coil ignition and also by removing all the gears from the gearbox and by taking the drive direct from the centrifugal clutch which is attached to the crankshaft. The space in the gearbox was then filled with polyurethane foam insulation to reduce its volume and hence the surface area available for oil cooling, to reduce churning caused by windage and to help retain heat. The piston skirt was shortened to reduce reciprocating mass and friction. Apart from these modifications the engine internals were unaltered, the standard camshaft timing and ignition timing being retained.

Externally the engine and gearbox casing were heavily insulated with rock wool, polyurethane foam and fibreglass tape. Prior to applying this insulation the cooling fins on both the cylinder barrel and cylinder head were filled in with 'plastic padding' to form a heat sink.

A thermocouple was fitted into the cylinder wall on the anti-thrust side to measure operating temperature. Because of the mode of engine operation required to obtain maximum economy, the engine never runs long enough to reach ideal working temperature, and thus any heat generated must be retained during the long coasting periods. In practice a mean cylinder wall temperature of 120 °C was maintained during the event.

12.7.4 Carburation

Fuel is metered into the engine via a special hand-made downdraught carburretter mounted on an extension to lift it clear of the cylinder head insulation. This carburetter always runs at WOT since it has no throttle plate or slide. It is not fitted with an idle or progression jet system. Fuel is sprayed into the Venturi, which is removable to permit the optimum size to be selected, via a 1/8 in copper tube. Fuel and air supply to the emulsion tube are both controlled by needle valves giving a very accurate control of mixture strength. The correct fuel level in the carburetter is obtained by adjusting the height of the 75 cm^3 glass fuel tank vertically prior to each acceleration, any decrease in height during the acceleration tending to weaken off the mixture. This adjustment also allows a lower fuel level in the carburetter to be used again helping weaker operation. The engine was started at WOT, a mixture setting being chosen to give the weakest possible mixture without misfiring at any condition during the full-throttle acceleration.

12.7.5 Ignition

Two 6 V motor cycle batteries connected in parallel were used to power a simple contact breaker and coil ignition system and also the starter motor. The original cam and contact breaker equipment was used with the appropriate condenser for coil ignition. Since the cam is attached to the crankshaft the plug is fired twice per cycle, i.e. on the compression stroke and on the exhaust stroke. Ignition timing was fixed at 35° before top dead centre (btdc).

12.7.6 Transmission

A thirteen-tooth bicycle chain sprocket attached to the inside of the centrifugal clutch drives a large sixty-four-tooth bicycle chain wheel mounted on a jackshaft above the gearbox casing via a standard bicycle chain. To the back of this chain wheel is attached another small (fifteen-tooth) sprocket which in turn drives a sixty-tooth sprocket which is attached to the free wheel unit on the back wheel. The primary drive is enclosed in a chain case. Total reduction is therefore 19.6:1 giving 3.93 (mile/h)/(1000 rev/min).

12.7.7 Weight

The total weight of the vehicle was 111 lb unladen, with the driver on board this increased to 265 lb.

12.7.8 Driving Technique

Since the centrifugal clutch engages at about 2500 rev/min and since the engine produces maximum torque at about 4500 rev/min, full-throttle accelerations between these two speeds are likely to give maximum economy. Obviously minimum sfc will not be available throughout this speed range, but at the top end of this range the engine appears to be operating most efficiently, the gearing being chosen to avoid running at a maximum engine speed in excess of the minimum sfc maximum torque range.

Because the Appleton test track is not level and incorporates several gradients, the duration of each acceleration is different and ranges from 5 to 7 s in conditions of low windspeed. Similarly the coasting periods vary, the longest on the downhill straight being as long as 5 min. In conditions of low windspeed, ten accelerations are required per lap which is completed in 11 min 24 s to average 10 mile/h.

At the startline for the event which is on a downward incline, the driver releases the brake, and the vehicle coasts for about 2 min during which time the driver adjusts the fuel level and engages the starter dog which is attached to a 12 V starter motor (from a Ford Escort). Operating this starter at only 6 V ensures that it cannot spin the engine fast enough to engage the centrifugal clutch.

When the speed drops to about 3 mile/h, the steering twist grip is turned which switches on the ignition via a micro switch in the twist grip handle, and the engine is started. As soon as the engine fires, the starter is automatically disengaged, and the vehicle is accelerated until the engine speed reaches

4500 rev/min as indicated by an electronic tachometer attached to the steering tiller. At this speed the main jet is rapidly closed via a small handle, and the ignition is switched off; the vehicle then coasts down to the starting speed.

The driver has to arrange his acceleration periods so that the high-speed periods occur on the straight parts of the track, thus avoiding any tyre scuffing on the corners. Increases in windspeed require changes in the driving technique and can increase the number of accelerations required per lap from ten to twelve.

The driver can judge his performance as he drives along by observing the level of the fuel in the glass tank, the anti-surge baffles giving a guide to the rate of fuel consumed.

In the 1976 event 39.85 cm^3 of fuel were used, giving a fuel consumption of 1141 mile/gal or 0.248 l/100 km. This value broke the challenging 1000 mile/gal target by 14.1%.

References

1. R. J. Greenshields. 150 miles per gallon is possible. *Soc. Automot. Eng. J.*, **58** (March 1950) 34
2. D. C. Carlson and H. D. Millay. Mileage marathon from 50 to 244 mpg. Paper presented at *Soc. Automot. Eng. St. Louis Sect. Meet., December 1969*, No. 700532
3. D. L. Berry. Nine ways to get better fuel mileage. Paper presented at *Soc. Automot. Eng. West Coast Meet., Anaheim, Calif., August 1974*, No. 740620
4. Mike McCarthy. Preposterous petrol parers. *Motor* (22 November 1975) 36–9
5. W. B. Crum and R. G. McNall. Effects of tire rolling resistance on vehicle fuel economy. *Tire Sci. Technol.*, **3**, No. 1 (February 1975) 3
6. Sir Harry Ricardo and J. G. G. Hempson. *The High-speed Internal Combustion Engine,* 5th edn, Blackie, London and Glasgow (1968), 130–41

Appendix A Glossary of Terms

Fuel economy is usually taken to refer to the number of miles a vehicle can be driven per unit volume of gasoline consumed (miles per gallon, or kilometres per litre), as measured over a driving pattern that should properly be closely defined. It should be noted that increased fuel economy means improved fuel use, i.e. more miles per gallon or kilometres per litre.

Mileage is synonymous with fuel economy in the US, though of course in units of miles per US gallon. Mileage is a dangerous term to use, for elsewhere it can simply mean distance travelled.

Fuel consumption of a vehicle usually refers to volumetric amount of gasoline used per distance driven (normally in units of litres per 100 km). It should be noted that it is essentially the inverse of fuel economy* but in different units and that increased fuel consumption means worse fuel economy, i.e. fewer miles per gallon.

Fuel consumption of an engine is related to steady-speed test bed performance and is usually quoted as a specific fuel consumption (sfc) expressed in gravimetric terms (pounds of fuel per horsepower hour in traditional units, kilograms per kilowatt hour in metric units or kilograms per megajoule in SI units). The work done by the engine can be that measured on the dynamometer brake (bsfc) or that measured inside the cylinder by an indicator diagram (isfc). The latter is the most basic engine characteristic since it omits all the pumping and frictional losses.

Fuel consumption can also be loosely used to describe merely the measured quantity of fuel consumed.

*Two points of importance emerge concerning these definitions, which must be used with caution.

(a) A percentage change quoted in terms of fuel economy will not be the same when quoted in terms of fuel consumption, i.e. a 20% increase in fuel economy, from say 20 to 24 mile/gal, corresponds to a 16.7% decrease in fuel consumption.

(b) There is a danger in taking the simple arithmetic means of fuel economy. Thus three measurements of say 10, 20 and 30 mile/gal, measured over the same distance, have an arithmetic mean of 20 mile/gal. However, the real average in terms of total distance driven divided by total fuel used is 16.4 mile/gal and is obtained by using the harmonic mean, i.e.

$$3\Big/\left(\frac{1}{10}+\frac{1}{20}+\frac{1}{30}\right)=\frac{3}{0.183}=16.4$$

Thermal efficiency of an engine is the basic engine work output characteristic compared with the most basic fuel input characteristic, namely its heat of combustion (or heating value or calorific value). It is unitless and usually expressed as a percentage. Brake thermal efficiency (bthe) and indicated thermal efficiency (ithe) correspond to bsfc and isfc, respectively, but it should be noted that increases in thermal efficiency are equivalent to decreases in sfc. Thermal efficiency can be applied (though not easily) to a vehicle driven over a defined pattern on the road or on a dynamometer. In this case, the thermal efficiency is the product of the engine's bthe and the mechanical efficiency of the drive train and the tyres. Thermal efficiency, properly defined from the various options just mentioned, is very useful for comparing radically different fuels.

Heat of combustion (variously referred to as heating value, heat content, heat value, calorific value, chemical energy and enthalpy of combustion) is the thermodynamic property of the fuel that defines its ability to release heat. It is usually the heat of combustion of the fuel as a room-temperature liquid at constant pressure and in a stoichiometric amount of air, assuming the final products to be water vapour and carbon dioxide, also at room temperature. It is worth noting that internal combustion engine technology deals exclusively with the net (sometimes called lower) heat of combustion. This should be clearly distinguished from the gross (or higher) heat of combustion which is measured when the products of combustion are allowed to equilibrate at room temperature, i.e. if water vapour condenses to liquid water, thereby releasing its latent heat. This preference is very reasonable, as water is in practice entirely in the vapour phase at the end of the working stroke. Other measures of heat of combustion are possible, which differ by small but significant amounts (see textbooks for further discussion).

Driving cycle is the non-steady-running driving pattern that can be imposed on a vehicle on the road or on the chassis dynamometer (or with more difficulty, and rarely, on the engine bed since often it is not equipped with the necessary inertia simulation and gearbox). Visual or audio instructions are given to the driver (or, in more sophisticated cases, electronic instructions are given to an 'automatic driver') to follow a speed–time or speed–distance pattern and to choose the relevant gears. Many different cycles have been devised to cover a wide range of driving conditions, e.g. urban conditions are simulated by the ECE 15 cycle in Europe or by the LA-4 cycle in the US and non-urban conditions by the EPA highway cycle in the US and even (though less well) by steady-speed tests at 90 km/h and/or 120 km/h in Europe.

Cold start refers to tests carried out with a cold engine or vehicles from specified ambient temperature. The word 'cold' has come into common use both to describe the emission test procedure with an ambient temperature of 15–30 °C as well as to refer to climatically cold conditions such as 0 °C or below. Care should be taken to distinguish between these two starting-temperature regimes.

Hot start refers to tests carried out with an initially hot engine, but here too care is needed to define properly the state of the engine and the fuel and lubricant temperatures, as well as the ambient temperature of the test, since these can all significantly affect measured fuel economy.

Appendix B Some Statistical Terms Commonly Used in Connection with the Measurement of Fuel Economy

Spread or range is a general term describing the spread or range of individual values (or of the means or standard deviations or other measures), normally using the maximum and minimum as the two extremes.

Precision is a statistical term describing the spread or degree of dispersion of values, often quantified in terms of the standard deviation of the sample or population of values.

Systematic error is the non-random error in the readings resulting from a consistent error (or bias) in the measurement apparatus.

Accuracy is a statistical term describing the closeness of the measured mean to the true mean of the whole population of values, often quantified in terms of the standard error of the mean (where the standard error of the mean is the standard deviation of the sample divided by the square root of the number of measurements). This statistical term cannot allow for systematic measurement errors.

Repeatability is a term referring to the degree of dispersion or distribution of values (i.e. the precision) measured in closely similar experiments (i.e. same operator, same day, same laboratory, etc.). It has also been given a particular quantitative definition which is commonly used: the difference between two values that will not be exceeded more than 5% of the time (approximately 2.8 multiplied by the standard deviation).

Reproducibility is the same as repeatability except that a greater difference in the experimental circumstances is usually meant (i.e. different operator, day or laboratory).

Standard deviation(s) of the sample is the root mean square value of the difference between the n measured values (x) and the sample mean ($\bar{x}$):

$$s = \left\{ \frac{\Sigma(x - \bar{x})^2}{n} \right\}^{1/2}$$

The % standard deviation is given by

$$\frac{s}{\bar{x}} \times 100$$

Standard deviation (σ) of the population is best given by

$$\sigma = s\left(\frac{n}{n-1}\right)^{1/2}$$

Standard error of the sample mean is given by

$$\sigma/n^{1/2}$$

Residual standard error (deviation) is what is left of the standard deviation of the sample after several independent variables have been allowed for in a model regression equation.

Appendix C A Guide to Mechanical Devices for the Improvement of Fuel Economy

It is not possible to cover the plethora of fuel-saving inventions individually. The patent literature over the last fifty years is replete with devices, the claims for which are frequently far in excess of their ability to perform, and many inventors' devices have disappeared as fast as they arrived.

However, there are some that do give real effects, though not always straightforwardly. The following guidelines are offered to help the reader give some semblance of order to this ongoing activity.

(a) The device may be entirely bogus.

(b) It may (and often does) cost more than the fuel it will save.

(c) It may (and usually will) alter the tune of the engine: it may exchange power for economy (e.g. mixture weakeners); it may exchange noise for economy (e.g. straight-through silencers); it may exchange emissions for economy (e.g. spark advance devices).

(d) It may affect the way the car is driven (but giving, of course, a valid economy gain): compulsorily (e.g. throttle dampers); optionally (e.g. vacuum guages, speed alarms).

(e) It may affect the heat balance of the engine (e.g. electric fans, radiator blinds).

(f) It may restore a worn or off-specification part that could perhaps be equally well replaced with a standard item (e.g. ignition boosting systems).

(g) It may have considerable deleterious side effects (e.g. mixture weakeners will seriously harm driveability and startability; heat balance changes may affect knock or run-on).

(h) It may make a genuine change in a feature of the engine, but one which does not affect fuel economy significantly (e.g. fuel atomizers improve mixture quality, but unless the mixture strength is weakened no fuel economy benefit can be seen).

(i) It might indeed be what all others have failed to understand and what the industry thus far has failed to produce: something that gives a genuine increase in economy over that of a well-functioning engine without prejudice to any aspect of performance.

Appendix D A Guide to Fuel-additive Inventions for the Improvement of Fuel Economy

Claims by inventors for fuel additives that can improve economy are in one sense easily made, for frequently the inventor has not done the very expensive work to evaluate the performance of his product. Even if he has done such work for one vehicle under certain conditions, he usually has not done so for a wider range of vehicles over high mileages, nor has he usually begun to assess any side effects, good or bad.

The following guidelines are offered to help put this difficult topic into perspective and should be read against the more thorough technical discussion in chapter 5.

(a) The product may be entirely bogus.

(b) It will, if it is valid, probably result in a small benefit (say 5% at most) compared with benefits from mechanical modifications.

(c) It may depend for its action on the effective weakening of mixture strength (e.g. thickeners, fuel diluters such as water, stoichiometric-point shifters such as blending components containing oxygen or nitrogen). Such a mixture weakening can be normally carried out more effectively by mechanical means.

(d) It may maintain the engine inlet system in a clean condition or even clean up a dirty one (see chapter 5).

(e) It may control malfunctions in the combustion chamber (e.g. spark plug fouling) or inlet system (e.g. carburetter icing) (see chapter 5).

(f) It may have significantly deleterious side effects: in the short term (driveability, octane quality, power output, exhaust emissions, fuel storage and handling); in the long term (engine cleanliness, octane requirement increase, corrosion).

(g) It may really be a *power*-enhancing additive (e.g. methanol); loose talk about 'engine efficiency improvement' may lead to the impression that fuel economy is improving also, whereas power and efficiency are not necessarily related and indeed often are inversely related.

(h) It may be claimed to give 'smoother combustion', but such flame-controlling additives are most unlikely to be found for spark ignition engines. Diesel fuel additives are known but will almost certainly be pro-knock in a gasoline engine and usually strongly so.

(i) It might indeed give a real benefit by some as yet poorly realized or understood mechanism.

Appendix E Why the Diesel Engine and the Lean-burn Gasoline Engine give Improved Fuel Economy Compared with the Conventional Gasoline Engine

D. R. BLACKMORE and R. BURT

Currently the question of near-alternatives to the gasoline engine as we know it today is being much discussed. The need for improved fuel economy has brought this question back into consideration and has somewhat replaced the discussion of more radical alternatives that the threat of tight emission control legislation had brought about. In this new context, the merits of the diesel engine are being seriously considered as an automotive power unit with greatly enhanced fuel economy. However, the discussion is clouded a little by the fact that so often the comparison between one power unit and another is made on only one of many possible bases (i.e. the two engines are compared at equal swept volume, surely one of the least technically and commercially significant bases). We therefore present here a brief summary of the fundamental reasons why the diesel engine (and the lean-burn gasoline engine) gives an enhanced fuel economy over the conventional gasoline engine.

The most significant operational area of difference occurs for the part-load condition. We shall therefore consider engines whose maximum power is the same, implying for the diesel and lean-burn gasoline engines a larger swept volume (and therefore larger and heavier engines) and smaller brake mean effective pressure (bmep) values. We take first the diesel engine.

For engines at full power, there is very little if any difference in brake thermal efficiency (bthe) (fuel economy) between the gasoline and the diesel engine. The higher compression ratios of the diesel (about 17:1 for direct injection and about 20:1 for indirect injection) compared with those of the gasoline engine (about 9:1) lead to some theoretical gain in cycle efficiency (see chapter 2), but these are largely offset by the increased friction that the high-compression engine design necessitates. There appear to be no significant differences attributable to the speed of the combustion process, but the turbulence and the radiant nature of the flame probably account for the increased heat rejection of the diesel to the combustion chamber and cylinder walls (figure E.1)[1] which further degrades the efficiency towards the gasoline engine value.

As these engines reduce power at a given speed, some significant differences begin to emerge. It must be remembered that the gasoline engine

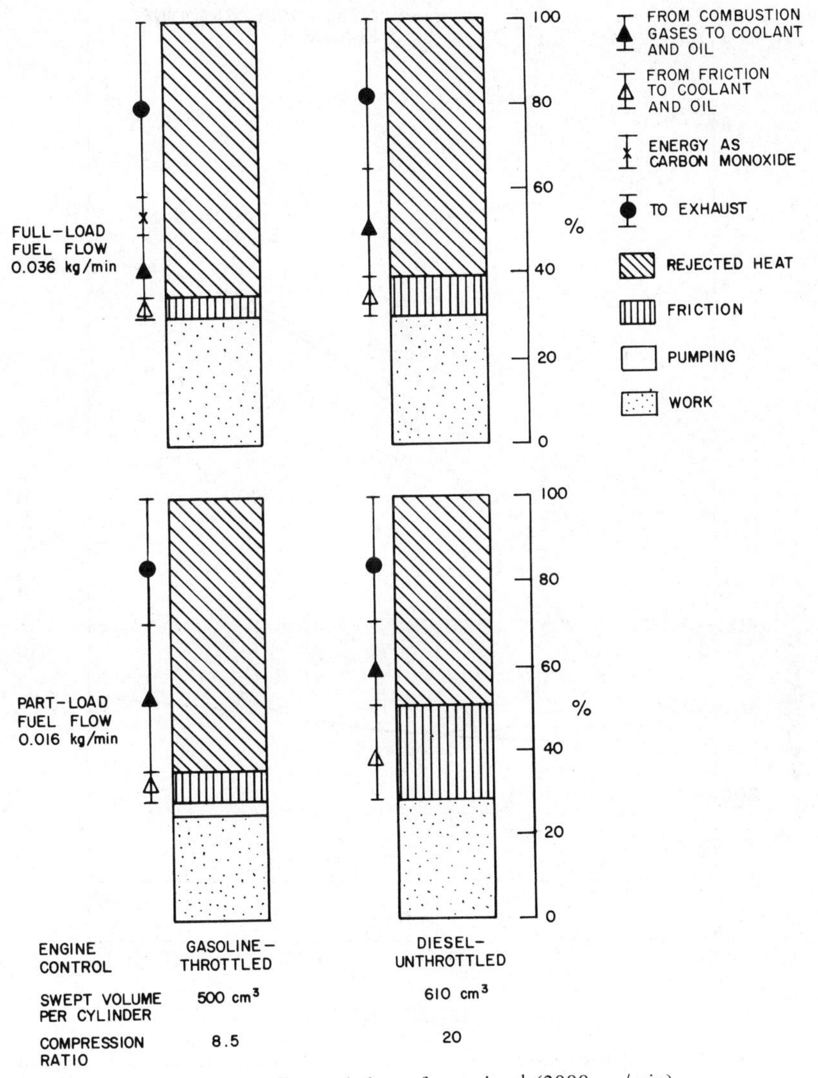

Figure E.1 Energy balance for engines[1] (2000 rev/min)

reduces power by throttling of a notionally constant air/fuel ratio mixture, whereas the diesel does so by reducing the fuel flow (and therefore increasing the air/fuel ratio) without any throttling action at all. The lean-burn gasoline engine occupies a position intermediate between these two. The consequences of power reduction by these different methods are as follows.

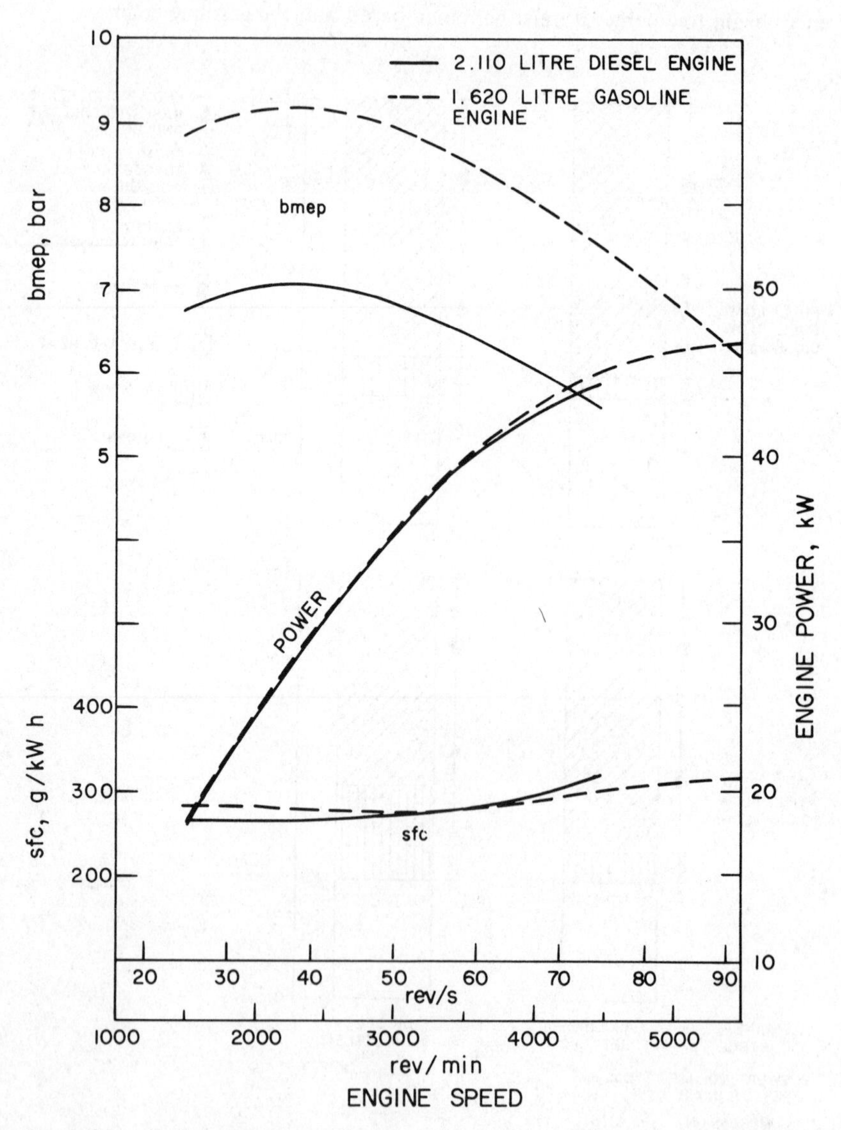

Figure E.2 Performance curves of comparable gasoline and diesel automotive engines[2] (1 bar = 10^5 N/m^2; 1 g/kW h = 2.78 x 10^{-2} kg/MJ)

(1) The leaner air/fuel ratio of the diesel gives rise to an increase in specific heat ratio γ, and so the cycle efficiency improves, moving towards the theoretical air cycle efficiency as the fuel supply is reduced towards zero.

(2) The throttling of the gasoline engine introduces pumping losses since work must be done to pump air from low inlet manifold pressures to much higher exhaust pressures: these losses can become a very large fraction of the fuel energy supplied for the light-load conditions so prevalent in everyday driving. The diesel does not suffer in this way.

(3) Heat losses with the diesel diminish as the power is reduced. In the diesel the air/fuel ratio increases as the power decreases, thus causing a decrease in cycle temperatures. In the gasoline engine these cycle temperatures are nearly constant whatever the throttle position, and the fall-off in heat losses is less marked. These lower part-load cycle temperatures with the diesel result in lower rejected heat from the combustion gases to the combustion chamber and cylinder walls (i.e. to the coolant and oil[1], see figure E.1).

There are therefore three thermodynamic factors, all working in the same direction, which cause the part-throttle bthe of the diesel engine to be greater than that of the conventional gasoline engine.

The combustion behaviour of the different engines at part throttle constitutes another area of difference, which is smaller, more complex and less well known than the above three major areas. These differences show themselves in the following ways.

(a) A significant exhaust CO concentration is a feature of the conventional gasoline engine and must represent a loss of chemical energy that does not occur in the diesel engine (figure E.1).

(b) The rate of heat release in the diesel actually increases as power is reduced, owing to the more ready availability of oxygen to the fuel in the turbulent diffusion flame. This is another reason why the combustion efficiency at part load is greater than that at full power. There is no comparable effect in the gasoline engine because the flame is essentially pre-mixed under all conditions.

(c) Subsequent use of the choke is required to achieve a combustible mixture in the gasoline engine during cold-starting and warm-up. No such overenrichment is required in the diesel, and a substantial fuel saving must result under these conditions.

A practical comparison of two engines (a 2.11 l Peugeot diesel and a 1.62 l Peugeot gasoline engine) on a constant-power basis has been given recently[2] and is shown in figures E.2 and E.3. The improvement in gravimetric brake specific fuel consumption (bsfc) for the diesel at low speeds and loads is evident from these figures and ranges up to 25% in magnitude (in volumetric terms the gains are even greater).

The lean-burn gasoline engine, on which much development work is in progress in various countries, provides an interesting though less well documented case.

Because ways of achieving power modulation by controlling mixture strength rather than by throttling are beginning to be found, all three of the thermodynamic factors listed above will also apply in some measure to the lean-burn gasoline engine, giving improved fuel economy. In addition, of the combustion factors, it is clear that the low CO emissions will minimize the loss of chemical energy. How far such power modulation will prove to be possible in practice is the current challenge for research, but, if significant success is achieved, then the engine will probably have advantages over the diesel engine in terms of maximum power, friction, noise, odour, weight and

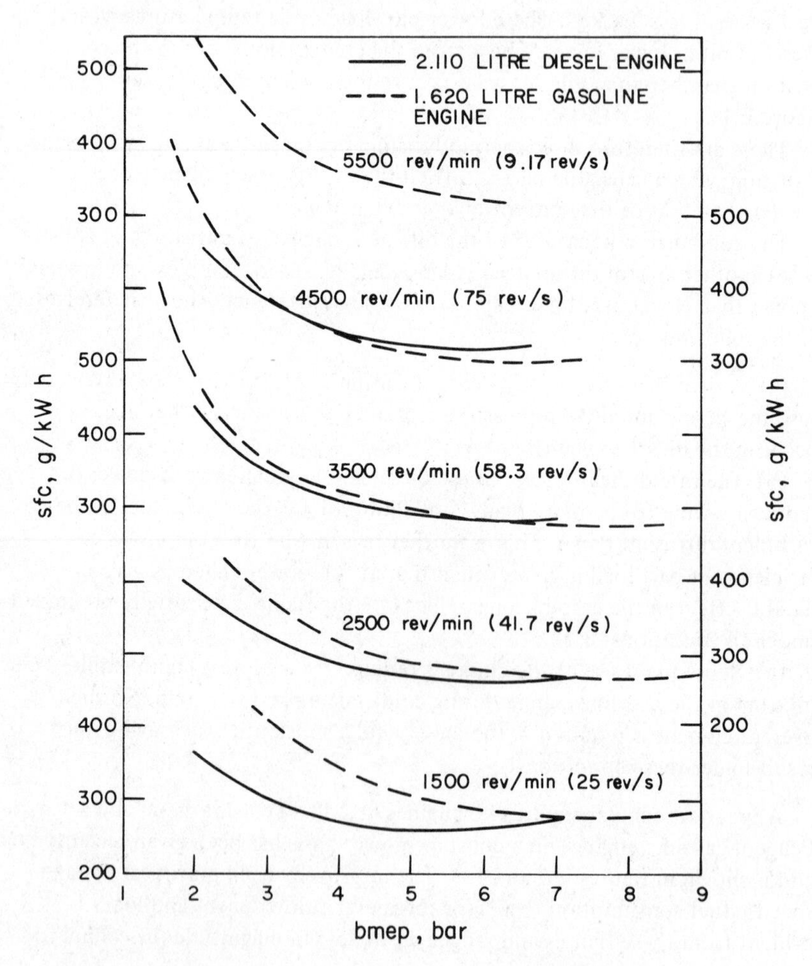

Figure E.3 Specific fuel consumptions (sfcs) of comparable gasoline and diesel automotive engines[2] (1 g/kW h = 2.78×10^{-2} kg/MJ; 1 bar = 10^5 N/m^2)

cost and will therefore be a very strong contender for the automotive market of the future.

References

1. *A Study of Stratified Charge for Light-duty Power Plants*, Vol. 1, contract for EPA 460/3-74-011/A, Ricardo Consulting Engineers, Shoreham-by-Sea (1927)
2. H. W. Barnes-Moss and W. M. Scott. The high-speed diesel engine for passenger cars. Paper presented at *January 1975, Conf. on Power Plants and Future Fuels, Inst. Mech. Eng.* No. C15/75

Appendix F Abbreviations, Units and Conversions

F.1 Abbreviations

A	car frontal area
API	American Petroleum Institute
ASD	a proprietary detergent additive marketed by Shell companies
ASTM	American Society for Testing and Materials
ASTM distillation	method of characterizing volatility by defining the %v of fuel evaporated as a function of temperature under specified distillation conditions (D-18 or IP-123 method)
ATF	automatic transmission fluid
bhp	brake horsepower
bmep	brake mean effective pressure
bsfc	brake specific fuel consumption
btdc	before top dead centre
bthe	brake thermal efficiency
C_p	specific heat at constant pressure
C_v	specific heat at constant volume
C_D	coefficient of air drag
C_R	coefficient of rolling resistance
CA	crank angle
CCS	controlled combustion system
CID	cubic inch displacement of an engine (engine displacement)
CT	closed throttle
CVCC	Honda compound vortex controlled combustion
CVS	constant-volume sampling
CVS-C	cold-start CVS test with 1972 UDDS cycle
CVS-CH	as above but with first 505 s of UDDS cycle repeated
CVS-H	hot-start CVS test with 1972 UDDS cycle
d	fuel density
DOT	US Department of Transportation
DPG	dipropylene glycol
ECE	Economic Commission for Europe (a UN body)
ECE 15	the official Economic Commission for Europe test cycle

EEC	European Economic Community (the Common Market)
EFE	early fuel evaporation
EFI	electronic fuel injection
EGR	exhaust gas recirculation
EPA	the US Environmental Protection Agency
ESA	electronic spark advance
E70T	%v of fuel evaporated at 70 °C in the ASTM distillation
E120T	%v of fuel evaporated at 120 °C in the ASTM distillation
f	residual gas fraction
fhp	friction horsepower
fmep	friction mean effective pressure
F	fuel/air weight ratio
F_c	stoichiometric (or chemically correct) fuel/air weight ratio
F_R	relative fuel/air ratio ($=F/F_c$), sometimes referred to as mixture strength or fuel equivalence ratio. This definition causes F_R to exceed unity as the mixture becomes fuel-richer. (N.B. Occasionally in the literature the term 'equivalence ratio' is ambiguously used to refer to the inverse of F_R. Also λ value or strictly air equivalence ratio is in use in some continental countries and is the exact inverse of F_R)
FID	flame ionization detection (of HC emission)
FT	full throttle
F-310	Chevron's proprietory detergent additive
G	road gradient
GRPA	Groupement des Rapporteurs de la Pollution de l'Air (a European group working on test procedures and reporting to the ECE)
h	humidity, fractional
hp	horsepower
HC	hydrocarbon
ihp	indicated horsepower
imep	indicated mean effective pressure
isfc	indicated specific fuel consumption
ithe	indicated thermal efficiency
ICA	ignition control additive
IW	inertia weight (usually as set on a dynamometer)
LA-4	the driving cycle used in the first part of the US emission test cycle (obtained from Los Angeles traffic data)
LNG	liquefied natural gas
LPG	liquefied petroleum gas
mep	mean effective pressure
mpg	fuel economy in miles per gallon

M_f	mass flow rate
MBT	minimum (spark advance) for best torque
MON	motor octane number
N/V	drive-train ratio of engine speed N, rev/min, to road speed V, mile/h
NAS	US National Academy of Sciences
NDIR	non-dispersive infra-red (measurement method for certain exhaust emissions)
OCAT	oxidizing catalyst
p_e	exhaust manifold pressure
p_i	inlet manifold pressure
php	pumping horsepower
pmep	pumping mean effective pressure
P_1, P_2, P_3, etc.	pressure at various points on an engine's $p-V$ diagram (see figure 2.2)
P_b	brake power
P_i	indicated power
PCV	positive crankcase ventilation
PEGR	proportional exhaust gas recirculation
PLU	Pierburg Luftfahrtgerate Union GmbH, manufacturer of a fuel flow meter
PROCO	Ford programmed combustion system
Q'	heat added per mass of gas
Q_c	gravimetric heat of combustion of gasoline
Q_v or Q_{cv}	volumetric heat of combustion of gasoline
r	compression ratio (or expansion ratio)
rahp	rubbing friction and accessories horsepower
ramep	rubbing friction and accessories mean effective pressure
RL	road load
RON	research octane number
RVP	Reid vapour pressure
sfc	specific fuel consumption
SAE	Society of Automotive Engineers (US professional body)
SG	specific gravity
SUS	Saybolt Universal Seconds (a viscosity measure)
T_i	inlet temperature
TCCS	Texaco controlled combustion system
TRC	Thornton Research Centre
T50E	temperature, °C, for 50%v of fuel evaporated in ASTM distillation
UDDS	the US urban driving dynamometer schedule
v	vehicle speed
VI	viscosity index
W	gross vehicle weight

WOT	wide-open throttle
γ	specific heat ratio C_p/C_v
η_b, η_i, η_n	brake, indicated, net thermal efficiency
ρ_f	density of gasoline

F.2 Units and Conversions

Density	1 lb/ft^3 = 0.01602 g/cm^3
	1 lb/gal = 0.09978 g/cm^3
Energy (heat or work)	1 Btu = 1.0544 kJ
	= 0.2928 W h
	1 ft lb = 1.356 J
	1 hp h = 2546 Btu
	= 745.7 W h
	= 2.685 MJ
	1 kW h = 3.6 MJ
Heat of combustion	1 Btu/lb = 2.3246 kJ/kg
Fuel consumption	1 gal/mile = 282.4 l/100 km
	1 US gal/mile = 235.2 l/100 km
Fuel economy	1 mile/gal = 0.3540 km/l
	= 72 furlong/firkin
	1 mile/US gal = 0.4251 km/l
Mean effective pressure (mep)	1 lb/in^2 (1 psi) = 0.07031 kg/cm^2
	= 6.895 kN/m^2 (kPa)
Power	1 hp = 0.7457 kW
Pressure	1 atm = 101.33 kN/m^2 (kPa)
	1 lb/in^2 (1 psi) = 6.895 kN/m^2 (kPa)
	1 bar = 105 N/m^2 (Pa)
Specific fuel consumption	1 lb/hp h = 0.6083 kg/kW h
	= 0.1690 kg/MJ
Speed	1 mile/h = 1.4667 ft/s
	= 1.6093 km/h
	= 0.4470 m/s
	1 ft/s = 0.3048 m/s
Torque	1 ft lb = 0.1383 kg m
	= 1.3552 N m
Tractive effort	1 lbf = 0.4536 kgf
	= 4.448 N
Viscosity	1 cSt = 1 x 10^{-6} m^2/s
	1 cP = 1 x 10^{-3} N s/m^2

Volume	1 gal = 1.2009 US gal = 4.546 l 1 US gal = 3.785 l 1 in^3 (cubic inch displacement (CID)) = 0.01693 l 1 l = 61.01 in^3

F.3 Conversions Between Commonly Used Units of Fuel Economy and Consumption

1 l/100 km ≡ 282.4 mile/gal ≡ 235.2 mile/US gal ≡ 100 km/l

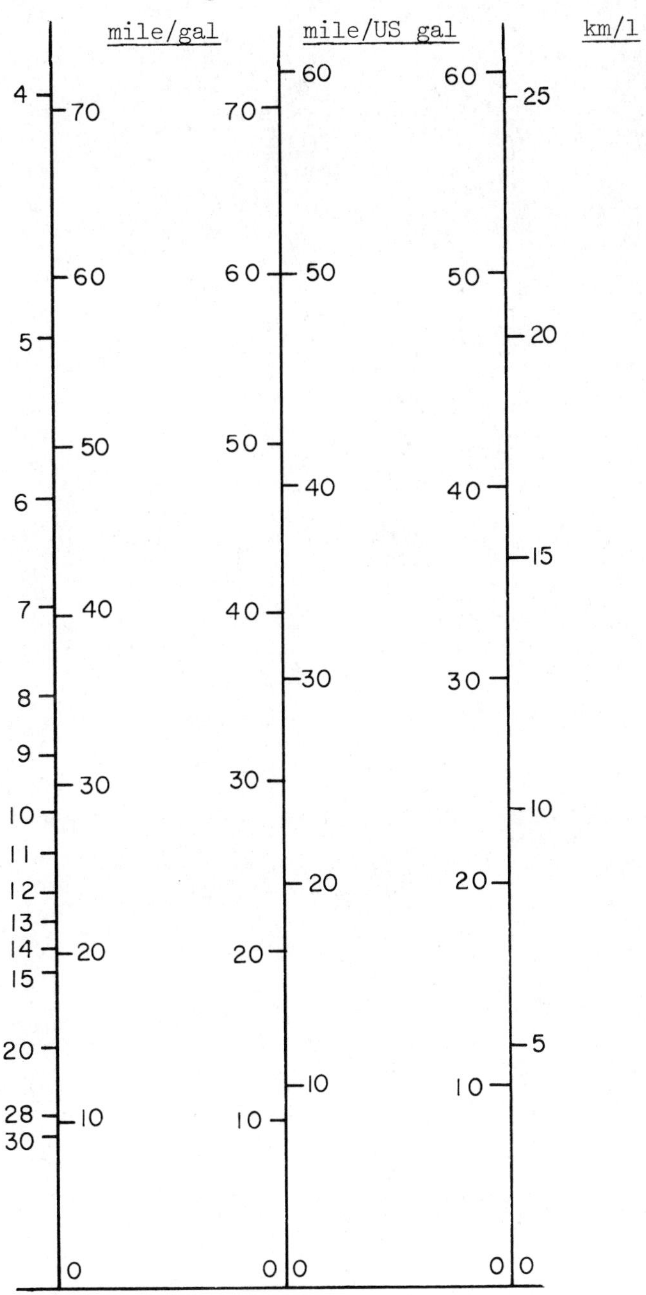

Index

Numbers in italics indicate main subject reference.